Relevance and Sustainability of Wild Plant Collection in NW South America

Grischa Brokamp

Relevance and Sustainability of Wild Plant Collection in NW South America

Insights from the Plant Families Arecaceae and Krameriaceae

Foreword by Prof. Dr. Maximilian Weigend
and Prof. Dr. Hartmut H. Hilger

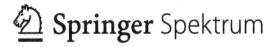 Springer Spektrum

Grischa Brokamp
Berlin, Germany

Dissertation Freie Universität Berlin, 2013

ISBN 978-3-658-08695-4   ISBN 978-3-658-08696-1 (eBook)
DOI 10.1007/978-3-658-08696-1

Library of Congress Control Number: 2015930519

Springer Spektrum

Printed on acid-free paper

Springer Spektrum is a brand of Springer Fachmedien Wiesbaden
Springer Fachmedien Wiesbaden is part of Springer Science+Business Media
(www.springer.com)

„A loucura é breve, longo é o arrependimento.“
Brazilian saying

Supervisor's Foreword

Nature's resources are being rapidly depleted as the need for space and resources of an ever growing number of humans increases. Plants, as primary producers, are the basis of nearly all natural productivity, but also the crucial biotic component in ecosystem services, such as carbon storage, and water and oxygen cycling. Man's impact on plant life – once limited and local – is now global and affects the entire biosphere. Hence, a growing number of scientists now state that we have entered a new geological epoch, the Anthropocene.

However, biotic resources are not unlimited and the ability to regenerate is often exceeded by the speed and extent of exploitation. It is in mankind's own vital interest to manage natural resources in a way that makes them last for future generations. This perspective of a "sustainability" depends greatly on various biological features and issues that can largely be captured by biological studies on growth, regeneration, productivity and recruitment. The associated social and economical facets are often less easily quantified and less predictable.

Palms are iconic for the tropics: beaches with coconut palms are the stereotype image of tropical paradise for people from the temperate zone. Extensive palm cultivation can have extremely negative side effects. Large-scale agricultural operations, though desirable, are among the ecologically most disruptive human activities. In respect to their effects on bio-diversity, there are probably few agricultural developments that are as devastating as large-scale oil palm plantations in the tropics. On the other hand, palms are suitable for large-scale cultivation operations under relatively natural conditions, and they can provide a vast range of products even in natural densities under sustainable harvest regimes.

Grischa Brokamp participated in the project "PALMS: Palm Harvest Impacts in Tropical Forests" funded by the EU Seventh Framework Programme. As a student researcher within the project's Work Package "Small Industries and Trade Based on Palm Products" he conducted his research at the Institute of Biology, Freie Universität Berlin, from 2009 to 2011 and at the Nees Institute for Biodiversity of Plants, Rheinische Friedrich-Wilhelms-Universität, from 2012 to 2013.

In the present study, Grischa Brokamp reviewed and analyzed the current extent of palm use in northwestern South America, providing interesting insights into associated mechanisms, their limitations and perspectives. He successfully implemented the work package's tasks, learned Spanish and got acquainted with research tools commonly used in economics. One of the most challenging tasks was the collection of trade data by means of interviews with stakeholders along the value chains of the different major palm products that are commercialized in the study region. For this, he developed and stepwise modified a now well-established and standardized research protocol for the acquisition of detailed data on production and marketing networks of palm products, which he published in Spanish.

The study focuses on understanding the commercial relevance of palms and the relation to the current patterns of use and sustainability. There are conflicts between use and conservation. Current exploitation, trade, and utilization are not in line with industrial practices and needs in a world of perpetual human population growth. Administrative and policy failures can quickly thwart any progress made.

Conflicting uses are influenced by specific attributes of the value chains. Understanding the biology of any particular species can provide important insights into their possible sustainable management, as also demonstrated for the case of rathany (Krameria lappacea). With this dissertation, Grischa Brokamp presents valuable aspects of the usefulness, commercialization and possible sustainable use of different plant products from neotropical palms, as well as from a valuable Andean medicinal plant, based on a thorough understanding of the biological characteristics of the plants.

Bonn & Berlin, September 2014   Prof. Dr. Maximilian Weigend

Prof. Dr. Hartmut H. Hilger

Acknowledgements

First of all, I would like to thank Prof. Maximilian Weigend for supporting me at all stages of my studies and for his invaluable help in the development and realization of this dissertation. I owe special thanks also to Prof. Hartmut H. Hilger for his lasting support. Many thanks are due to all members of the institution, particularly to everybody I worked with, namely Dr. Federico Luebert, Dr. Tilo Henning, Dr. Markus Ackermann, Natalia Valderrama, Moritz Mittelbach, Michaela Achatz, Christian Schwarzer, and all those I have forgotten to mention here, albeit they might have supported me in various ways in the past decade.

I am especially thankful to Prof. Henrik Balslev, Dennis Pedersen, Prof. Anders Barfod, Henrik Borgtoft Pedersen, and all the others I worked with at Aarhus University or met on workshops and symposia in South America and elsewhere throughout my participation in the EU-FP7-PALMS project. Research on palms was funded by the FP7-PALMS-project, Palm Harvest Impacts in Tropical Forests, FP7-ENV-2007-I, contract no. 212631 (http://www.fp7-palms.org). Field data provided by Henrik Borgtoft Pedersen are gratefully acknowledged.

Many thanks to everyone who supported me in design, translation, and refinement of the standardized interview protocol, i.e., many thanks to the co-authors, to Lucia de la Torre, Daniela Baldassari, Renato Valencia, Daniela Cevallos, Janice Jacome, Narel Paniagua Zambrana, Erika Blacutt, Jaime Navarro, Zorayda Restrepo, Carlos Martel, Yakov Quinteros, all our brave interviewees from numerous communities, NGOs, and companies.

The studies on Krameria were carried out in the framework of a private-public-partnership for developing a standard for sustainable wild collection practices for Rhatany by WELEDA AG Arlesheim, Switzerland and GTZ – Gesellschaft für Technische Zusammenarbeit, Eschborn, Germany (Nachhaltige Wildsammlung, Biotop-Erhaltung und Kultivierung der gefährdeten Andenpflanze Ratanhia: VN 81060590). Many thanks to A. Ellenberger (WELEDA AG) for his support. Field data provided by Richard Aguilar Carazas (Arequipa), Tilo Henning, Oliver Mohr and Christian Schwarzer (Berlin) are gratefully acknowledged, as is the support of Lis Scherer in the preparation of the LM-sections.

Special thanks to Dr. Rommel Montúfar for his support in all aspects of fieldwork in Ecuador. Field data provided by Rommel Montúfar and Janice Jácome are gratefully acknowledged. Many thanks to César Augusto Grandez Rios for fieldwork support in Iquitos (Peru).

Many thanks, of course, to my family. I would not have been able to complete this study without the steady support of Hannah Gritz, to whom I dedicate this work.

Berlin, June 2013   Grischa Brokamp

Contents

14

List of figures

List of tables

1 Introduction

1.1 Motivation and Research goals

People depend on natural resources supplied by wild plants, for food, construction, energy, and medicine all over the world and particularly in developing countries (Pimentel and Pimentel, 2008). Apart from the direct use or consumption of wild plant resources, the commercialization of plant raw materials or the sale of products manufactured from them provide cash income, reduces poverty, and represents a safety net during emergencies and times of food shortages. Furthermore, human societies also depend on a variety of indirect ecosystem services, such as water catchment, erosion control, carbon storage, etc. (Balslev, 2011), a major portion of which is provided by wild plants (Bastian, 2013).

However, currently and in the decades to come several challenges are looming that pose a threat to entire ecosystems and by that to numerous wild plant populations, and the ecosystem services they provide, consequently affecting the welfare and sustenance of mankind:

(I) A growing global population, heading for nine billion by 2040, has to ensure sufficient availability of food, water and energy to meet future needs. This will definitely have a disproportionately negative impact on the environment (Ehrlich & Holdren, 1971). Already by 2030, the world population will consume 50% more food and 45% more energy, as was estimated by the United Nations (2012) and plants will play a major role in satisfying the increased demand of both in the future (e.g., Berndes *et al.*, 2003).

(II) All over the world and especially in the tropics, natural ecosystems are subject to intensive human impact and the conservation of plant resources they provide is directly dependent upon active management (Altieri *et al.*, 1987). Particularly tropical forests are degraded by logging and the overexploitation of wild plant resources other than wood or – even worse – are completely destroyed by slash-and-burn agriculture (Rudel and Roper, 1997). Clearly, this affects local and global biodiversity and often results in permanent changes of land use (deforestation), which in turn has an effect as a driver in climate change (Tinker et al. 1996).

(III) Climate change is expected to be a major driving force for ecosystem change in the decades to come (IPCC, 2001, 2007). Associated changes in temperature, precipitation, and seasonal variation represent an profound threat to biodiversity (Bastian, 2013) and also constitute a major challenge for nature conservation (Svenning & Sandel, 2013). Already 30 years ago, a significant effect of global warming was discernible in wild plant populations (Root *et al.*, 2003) causing shifts in species distribution and abundance (Parmesan & Yohe, 2003), which, among other factors, lead to an increased extinction risk of species (Thomas *et al.*, 2004). Therefore, an increasing loss of biodiversity can be expected through the effects of climate change alone, especially in regions with a high proportion of fragmented or isolated habitats (Rannow *et al.*, 2010) and species that are already threatened by changes in land use are particularly threatened (SMUL, 2008).

Overall, the impacts of both climate change and (increasingly) destructive human activities are closely connected and represent the most critical factors that creating new limits for our environment's resilience and ability to supply (Pasztor & Schroeder, 2012). Sadly, food shortage (or inappropriate distribution of produced food) and the resulting malnutrition as well as scarcity of drinking water already represent a huge problem for large parts of the world population, particularly in developing countries, which in 2010 resulted in around 925 million undernourished people worldwide (FAO, 2010).

NW South America represents a global hotspot of vascular plant biodiversity (Mutke & Barthlott, 2005) and hence there is an extremely high number of useful plant species to be found in the countries Bolivia, Colombia, Ecuador, and Peru, most of which are collected from the wild (De la Torre *et al.*, 2008; Reyes-García *et al.*, 2006; Aguirre *et al.*, 2002; Duivenvoorden *et al.*, 2001). However, legal and administrative frameworks that regulate the extraction and trade of non-timber forest products (NTFPs) in these countries are highly fragmented and inefficient; amounts of plant resources extracted from the wild are neither regularly controlled nor documented (De la Torre *et al.*, 2011).

These data are required and need to be assessed in order to understand the relative and absolute socio-economic importance of individual plant species and, thus, represent a crucial foundation to determine the value of corresponding ecosystems. "Lack of this understanding and failure of markets in reflecting the value of ecosystems mean that information conveyed to economic decision-makers at all levels remains incomplete. Typically, the full social and environmental benefit of these goods and services and the impact and full cost of their degradation are not translated in a way that will ensure optimal decisions for both the economy and the environment" (Newcome *et al.*, 2005). Welfare and sustenance of mankind in the decades to come therefore eminently depend on the success of establishing policies best suited to mitigate the combined impact of the main causal and intricately linked key factors for environmental degradation of ecosystems (i.e., increase of human population, climate change, and unsustainable management practices or destructive land use).

The present work encompasses topics that range from basic botanical research through to the economic botany of plants that are subject to commercial exploitation in NW South America. The results here presented come from several studies that were conducted in order to contribute to a better understanding of the current situation of plant species that are regularly harvested from the wild and commercialized. Both biological baseline data and data on socio-economic importance, extent of trade, and economic value of plant raw materials are provided and may act as background data required for the design and implementation of programs that foster the conservation and sustainable exploitation of corresponding species.

1.2 Species

1.2.1 Arecaceae

The palm family (Arecaceae or Palmae) represents a large and diverse plant family of mono-cotyledonous flowering plants. According to the latest classification the family is divided into the 5 subfamilies Calamoideae, Nypoideae, Coryphoideae, Ceroxyloideae, and Arecoideae (Asmussen *et al.*, 2006), which comprise 28 tribes, 27 subtribes, and around 2,400 species in 183 genera (Dransfield *et al.*, 2008; Govaerts *et al.*, 2013). Palms are predominantly found in tropical and subtropical regions of the world and a major portion of palm species thrives in tropical rain forest habitats. Some seasonal and semi-arid habitats are also relatively palm rich, and a couple of species also even occur as a characteristic components of some desert floras (Boyer, 1992; Dransfield *et al.*, 2008).

South America has 457 palm species in 50 genera (Pintaud *et al.*, 2008), whereas numer-ous tribes of the subfamily Arecoideae dominate the palm flora here, with only 3 genera (*Chamaedorea, Geonoma & Bactris*) accounting for one third of all American palm species (Henderson *et al.*, 1995; Dransfield *et al.*, 2008). However, Ceroxyloideae and Calamoideae are also of importance in South America, the latter primarily because of the high abundance of individuals of just seven species (e.g., *Mauritia flexuosa*) from the tribe Lepidocaryeae (Dransfield *et al.*, 2008). Systematic affinities of palm genera and species dealt with in this thesis are presented in Table 1.1.

Although a typical palm builds a solitary stem – a shoot with a single apical meristem bear-ing a crown of leaves – many palms deviate from this bauplan and develop clustering stems or form shrubs, or even lianas (Dransfield *et al.*, 2008, Tomlinson, 2006). Notably, palm stems neither produce a bark nor do they consist of true wood with annual rings. This reflects their monocotyledonous character: In contrast to many other trees, palm stems contain vas-cular bundles scattered throughout a softer parenchymatous tissue, which are most densely packed in the outer part and decrease in number towards the center of the stem. This results in the fact that the fibrous palm timber is completely different to timber that comes from non-palm tree species and is the cause of the enormous flexibility and rigor of palm stems (Parthasarathy & Klotz, 1976; Dransfield *et al.*, 2008). Additionally, many palms are very well adapted to grow in seasonally flooded areas that are not suited for agriculture, where they often develop dense and monotypic stands (e.g., *aguajales* = dense stands of *Mauritia flexuosa*; *taguales* = dense stands of *Phytelephas aequatorialis*; e.g., Prance, 1979; Schlüter *et al.*, 1993).

Table 1.1 Systematic affinities of palm genera and species dealt with

Subfamily	Tribe	Subtribe	Genus
Calamoideae	Lepidocaryeae	Mauritiinae	*Lepidocaryum tenue*
			Mauritia flexuosa
Ceroxyloideae	Ceroxyleae		*Ceroxylon* spp.
	Phytelepheae		*Aphandra natalia*
			Phytelephas spp.
Arecoideae	Iriarteeae		*Iriartea deltoidea*
			Socratea exorrhiza
			Wettinia spp.
	Cocoseae	Attaleinae	*Attalea* spp.
		Bactridinae	*Astrocaryum* spp.
			Bactris spp.
		Elaeidinae	*Elaeis* spp.
	Euterpeae		*Euterpe* spp.
			Prestoea acuminata
			Oenocarpus bataua
	Geonomateae		*Geonoma* spp.
	Leopoldinieae		*Leopoldinia piassaba*

Due to their high diversity, abundance and interactions, many palm species play key ecological roles and provide numerous ecosystem services (Johnson and the IUCN/SSC Palm Specialist Group, 1996). They are also of great cultural and economic significance (see, e.g., Endress *et al.*, 2013; Gilmore *et al.*, 2013; *Mauritia flexuosa* in lowland Peru), ranking third after grasses (Poaceae) and legumes (Fabaceae) in overall economic importance. According to Johnson (2011), palm products typically fall into three different general categories, which are (I) primary products, (II) secondary or by-products, and (III) salvage products. Primary products represent the chief commercial (or subsistence) product, secondary and salvage products refer to useful items or material directly generated by processing and harvesting of the primary product, respectively. Another categorization is based on the type and degree as well as on location and level of sophistication in the processing of palm products. (I) The majority of palm resources represent products for immediate use, which are extracted from the wild by means of an ax or machete and are exploited at subsistence levels only (palm heart for direct consumption, fruits, and fronds for thatch). (II) Production of goods that require a modest amount of processing, few tools, and which are produced in locations that are not exclusively designated for processing is refered to as cottage-level processing (traditional extraction of palm mesocarp oil, weaving of mats, manual carving of vegetable ivory). (III) Small-scale industrial processing implies the need for specialized equipment, a dedicated locality where processing takes place, and a number of skilled workers, that produce goods manually, semi-mechanized, or mechanized (Canning of palm hearts, distillation of palm wine). (IV) Large-scale industrial processing is distinguished from the preceding in terms of the greater physical size of the processing facility, a higher level of sophistication in the processing itself through more complicated mechanical devices and certain highly skilled workers to operate and maintain equipment (African palm oil factories, processing of most products with export quality; Johnson, 2011).

Palms (Arecaceae) stand out as a plant group of extraordinary usefulness and are of particular socio-economical importance on a daily basis for numerous rural communities in northwestern South America (e.g., Lévi-Strauss, 1952; Macía, 2004; Paniagua-Zambrana et al., 2007; Macia et al. 2011). However, the bulk of utilized native palm species is harvested or managed in wild populations in various ways of which some are sustainable and others are destructive (Balslev, 2011) and *Bactris gasipaes* represents the only exception that is fully domesticated (Johnson, 2011). Consequently, palm species used for subsistence purposes are principally locally depleted close to villages, while commercialised species are generally more widely depleted (Kvist & Nebel, 2001; Iquitos, Peru). Overall only few (old world) palm species represent cultivated major crops, i.e., coconut, date, and oil palm (Johnson, 2011). Therefore, palms are perfectly suited to act as object of study in research on overall importance, trade extent, and the impact through harvest of wild plant raw materials in subsistence and cash economies in the midst of a global hotspot of biodiversity. A case study on the productivity and management of *Phytelephas aequatorialis* was performed in order to investigate the link between production rates of raw materials under different regimes of management and abiotic factors such as altitude and exposure to sun light. Detailed information on *P. aequatorialis* is presented in Chapter 4.1.

1.2.2 *Krameria lappacea*

Krameria lappacea, a slow-growing shrub that shows intriguing ecological characteristics and is found in an extreme environment of seasonal aridity. It is subject to destructive harvest from the wild for commercialization (Weigend & Dostert, 2005). However, scientific baseline data is scarce and a deeper understanding of the biological function of this commercially exploited plant species is non-existent. Data on abundance and productivity of *Krameria* are absent from the scientific literature. Its ecological role and relevance for the associated ecosystem remain poorly understood. Details on the systematic background of the family Krameriaceae as well as on ecological aspects and on commercial uses of *Krameria lappacea* are presented in Chapter 5.1 and 6.1.

1.3 Ecosystem goods and services

1.3.1 What are ecosystem goods and services?

Ecosystems and their biological diversity offer a wealth of goods and services, providing mankind with essential basic supplies and represent the foundation for economic prosperity and other aspects of welfare (Newcome et al., 2005).

In a broad sense, the term ecosystem services refers to the range of characteristics and processes through which natural ecosystems, and the species that they contain, help sustain and fulfil human life (Daily, 1997). These services regulate the production of ecosystem goods, which refer to the natural products used by humans on a daily basis, such as wild fruit and nuts, forage, timber, game, natural fibres, spices, medicines and so on. Ecosystem goods thus represent the various products, i.e., the direct, economical value of an ecosystem and the associated biodiversity (Newcome et al., 2005).

More importantly, ecosystem services support life through the regulation of essential processes, such as the purification of air and water, the pollination of crops, nutrient cycling, decomposition of wastes, and generation and renewal of soils, as well as by moderating environmental conditions by stabilising climate, reducing the risk of extreme weather events, mitigating droughts and floods, and protecting soils from erosion (MEA, 2005).

Ecosystem services thus represent the indirect value of an ecosystem and since the release of the Millenium Ecosystem Assessment (MEA, 2005) the number of studies on the evaluation of ecosystem services has grown, each of them defining and subcategorizing ecosystem services in slightly different ways (Ojea et al., 2010). According to Newcome and collaborators (2005), ecosystem services can be grouped into the following six categories, which are broadly based on both their ecological and economic function: (I) Purification and Detoxification: filtration, purification and detoxification of air, water and soils; (II) Cycling Processes: nutrient cycling, nitrogen fixation, carbon sequestration, soil formation; (III) Regulation and Stabilisation: pest and disease control, climate regulation, mitigation of storms and floods, erosion control, regulation of rainfall and water supply; (IV) Habitat Provision: refuge for animals and plants, storehouse for genetic material; (V) Regeneration and Production: production of biomass providing raw materials and food, pollination and seed dispersal; and (VI) Information/Life-fulfilling: aesthetic, recreational, cultural and spiritual role, education and research. Clearly, plants are the crucial ecosystem component in the provision of the six categories mentioned above.

1.3.2 Importance and valuation of ecosystem goods and services

Establishing the link between a given ecosystem and its goods and services and how these are valued by individuals is the key to an understanding of the importance and value of ecosystems and their incorporation in economic and other policy decision-making (Newcome et al., 2005). This topic gave rise to a novel subfield of economics (environmental economics), which undertakes studies of the economic effects of national or local environmental policies and includes concepts such as market failure (unfettered markets fail to allocate resources efficiently) and valuation of the environment (assessment of the economic value of ecosystems; Harris, 2006; Hanley et al., 2007). A central concept of environmental economics represents the determination of total economic value (TEV), which primarily is composed of use values that involve some interaction with the resource, either directly or indirectly as explained in Chapter 1.3.1, but also takes non-use values into account. Non-use values are associated with benefits derived simply from the knowledge that the ecosystem is maintained and are, by definition, not associated with any use of the resource or tangible benefit derived from it. When goods and services are provided in actual markets, the price individuals pay is at least a lower-bound indicator of how much they are willing to pay for the benefits they derive from consuming that good or service. For environmental resources which are not traded in actual markets, such behavioural and market price data are missing. Regardless of whether all components of TEV can be expressed in monetary terms for a given ecosystem good or service, the concept is reported to be useful in gathering the necessary information for more sustainable decision-making (Harris, 2006; Hanley *et al.*, 2007; Newcome *et al.*, 2005).

According to Newcome and collaborators (2005) four factors need to be taken into account when the importance of ecosystem goods and services are incorporated in economic decisions: (I) Understanding of the ecological functions that produce ecosystem goods and services; (II) Interface ecology and economics, which involves identification of those goods and services that are directly supplied, indirectly provided or (positively or negatively) influenced by human activities; (III) Definition and quantification of the economic benefit provided by goods and services, taking account of the components of the total economic value that applies in each case; and (IV) Distribution of benefits that derive from ecosystem goods and services among different beneficiary groups (spatially defined at the very least) and time periods, i.e., identification of different stakeholders, which is also useful in understanding the distribution of the costs involved when ecosystems are degraded.

1.4 Ecosystem goods, legal extraction, and value chains

1.4.1 NTFPs and MAPs - Plant resources from nature

Before the 1980s, timber was perceived as the primary product obtained from forests and accordingly forest policy and formal management were focused on it, largely downplaying other available goods such as, e.g., mushrooms, resins, leaves, and fruit, while completely ignoring provided ecosystem services and conservation. These „other products" or non-timber forest products (NTFPs) were defined as "all the biological material (other than industrial round wood and derived sawn timber, wood chips, wood-based panel and pulp) that may be extracted from natural ecosystems, managed plantations, etc. and be utilised within the household, be marketed, or have social, cultural or religious significance" (Wickens,

1991). However, since then, mainly due to alarming rates of deforestation, awareness has increased that forests generate many secondary products and services. These ecosystem services that are essential to local communities and required by society at large must be accommodated by forest management (Belcher *et al.*, 2003; Mantau *et al.*, 2007). Many alternate terms that are used more or less as synonyms for these secondary products have been in use since then, with non timber forest products (NTFPs; De Beer & McDermott, 1989) and non wood forest products (NWFP; Chandrasekharan, 1995) as the most widely employed terms (Belcher, 2003). Despite subtle differences, these terms all refer to ecosystem goods such as mushrooms, fruit, leaves, plants and animals collected in forests and used as food, fodder, medicine, as raw materials for production of handicrafts or cultural objects and as a source of income and subsistence (FAO, 2005; Mantau *et al.*, 2007).

Medicinal and Aromatic Plants (MAPs) refer to a component or sub-sector of NTFPs. They comprise plants used to produce pharmaceuticals, dietary supplement, cosmetics and personal care products, as well as some products marketed in the specialty food sector (Pyakurel & Baniya, 2011). According to the World Health Organization, the majority of the world's human population, especially in developing countries, depends on traditional medicine based on MAPs (WHO 2002). Worldwide, some 50,000 to 70,000 plant species are known to be used in traditional and modern medicinal systems (Schippmann *et al.*, 2006). The majority of MAPs are traded locally, regionally, and nationally, but ca. 3,000 MAP species are traded internationally (Lange & Schippmann, 1997). Altogether, many different plant parts and resources are harvested as NTFPs, including roots, tubers, leaves, bark, twigs, branches, flowers, fruits, nuts, seeds, gums, saps, resins, latexes, and essential oils (Walter, 1998).

Due to the fact that most NTFPs do not require the cutting of trees and provide subsistence and income for local people, the support of use and commercialization of these products has been considered as a promising tool for achieving conservation objectives while at the same time supporting development and improving livelihoods of rural communities (Peters *et al.*, 1989; Falconer, 1990; Plotkin & Famolare, 1992; Nepstad & Schwartzman, 1992). However, an unconsidered promotion of NTFP commercialization based on extraction from the wild generally bears ecological and livelihood risks (Belcher & Schreckenberg, 2007) and in many cases is reported to rather lead to resource depletion. Therefore, Kusters and collaborators (2006) stated that NTFP trade is unlikely to reconcile development and the conservation of natural forest.

1.4.2 Legality of the extraction of and trade in NTFPs in NW South America

In 1992, the Convention on Biological Diversity (CBD), set up by the United Nations, was held in Rio de Janeiro due to a common concern about the significant reduction of biological diversity by certain human activities. Here, the contracting parties affirmed that individual states have sovereign rights over their own biological resources and that they are responsible for conserving their biodiversity by using their biological resources in a sustainable manner. Sustainable use was here defined as "[...] the use of components of biological diversity in a way and at a rate that does not lead to the long-term decline of biological diversity, thereby maintaining its potential to meet the needs and aspirations of present and future generations" (CBD, 1992).

In order to comply with this international agreement, individual states approved to translate these decisions into national law to ensure that activities in areas within their jurisdiction (in the case of components of biological diversity) and also beyond the limits of national jurisdiction (in the case of processes and activities) do not cause damage to the environment (CBD, 1992). Generally, this represents the legal background for sustainable use and conservation of biodiversity on the (inter)national level, which includes any biotic component of ecosystems with actual or potential use or value for humanity, such as NTFPs.

Two decades after the CBD, Bolivia, Colombia, Ecuador, and Peru have all implemented a national legislation that prohibits unsustainable use of forest resources. Although there are different procedures for legal harvest of NTFPs from public and private lands with commercial purposes, for the issuance of a permit, generally, a management plan has to be provided that verifies biological sustainability of resource extraction. Harvest of NTFPs for subsistence however, does not require any permit in Ecuador, Peru, and Bolivia; and albeit it requires a permit in Colombia, its issuance does not require proof of sustainable resource extraction (De la Torre *et al.*, 2011). As a consequence, since there is no or only insufficient control of extracted amounts and, particularly on the local level, a major portion of raw material trade proceeds informally only, a clear separation between subsistence use and commercial use is leveraged: Many local industries depend on raw materials from local markets and suppliers who in turn obtain raw materials from peasants or local communities that collect small quantities under the subsistence regulation (Bernal *et al.*, 2011).

1.4.3 Value chain analysis

A value chain, also known as supply chain, or market chain (Neumann & Hirsch, 2000), „[...] describes the full range of activities, which are required to bring a product or service from conception, through the different phases of production (involving a combination of physical transformation and the input of various producer services), delivery to final consumers, and final disposal after use", as defined by Kaplinsky & Morris (2002). Value chains of NTFPs thus comprise the total of proceeding activities broken down according to type and location of stakeholder activity, such as production and provision of resources, distribution and supply, processing, storage, transport, marketing, and sale. In this, the relative importance of each of these categories may differ among products, they may not occur sequentially and some may even be repeated or omitted (Marshall *et al.*, 2003). The term value chain highlights the value that is added through these different processes and activities of individual stakeholders (Schreckenberg *et al.*, 2006). The objectives of a value chain analysis are: (I) identification of the main actors or organizations in the commercialization chain from resource production through to the final consumer; (II) identification of specific stakeholder activities and share of benefits; (III) identification of different routes, current extent, and potential of production and commercialization regarding individual NTFPs; and (IV) assessment of the condition of the marketing chain (Marshall *et al.*, 2006). Accordingly, a value chain analysis provides qualitatitive and quantitative data which include the characterization of all segments, actors, products and market channels, as well as the quantification of invested time, distance, income, profit, and its variations along the segments.

Several ways of calculations are in use for a characterization and quantification of relative and absolute benefits obtained by individual stakeholders. The most common are calculation of gross income (difference between investments and sales price), commercialization margin (difference between investments and sales price divided by final consumer price), and share of consumer´s price (Price of the product at each segment divided by final consumer price; Marshall *et al.*, 2006). Data required for a comprehensive value chain analysis is usually obtained from official statistics and by performing interviews with stakeholders (of all segments). Value chains may be short and simple, particularly those of locally traded products, with primary producers, i.e., individual farmers or harvesters, selling their products directly to consumers. Value chains that span over larger geographical ranges and that involve a higher level of sophistication in processing tend to be more complex and the recent trend towards increased globalization has transformed the way business works, making value chains of NTFPs more complex and difficult to manage (Belcher & Schreckenberg, 2007). Calculations of the commercialization margins and the share of the consumer´s price may be challenging, particularly for resources and products that are processed or transformed or do not have a standard unit of measure when passing through the supply chain. Therefore, it is not always possible to present this type of analysis for every NTFP. Overall, a value analysis provides understanding of why things work the way they do and whether changes in governance or behaviour may be needed in order to ensure equitable distribution of benefits among stakeholders and to foster sustainability in production and trade of NTFPs (Marshall *et al.*, 2006).

1.5 Aims and scope of the study

1.5.1 Research questions

1. How can the data on trade with wild plant resources be obtained in a standardized manner?

2. What are the economically most important native palm species, raw materials and products in northwestern South America in terms of turnover and amounts traded?

3. What is the overall socio-economic importance of the trade in palm resources for primary producers and which share of the overall benefits do they obtain?

4. How are leaf and fruit production in *Phytelephas aequatorialis* correlated to environmental factors (altitude and exposure to sunlight) versus management?

5. What is the biology of *Krameria lappacea* and how can sustainable management be based on a biological understanding of the species?

1.5.2 Specific objectives

1. Develop and test a standardized research protocol to document marketing networks for raw materials and products that derive from native palm species across northwestern South America with standard questions to actors at different levels of marketing chains.

2. Gather available information concerning the commercial extraction and sale of native palm raw materials and products in northwestern South America, including scientific data, information from official statistics, and obtain data by means of interviews with stakeholders regarding harvest, consumption, and detailed trade information.

3. Compile and analyse available data on palm trade in order to provide an estimate on the relative and absolute economic importance of individual palm species and on the benefits and overall socio-economic importance of trade in native palm resources for primary producers.

4. Determine and compare leaf and fruit productivity of *Phytelephas aequatorialis* individuals in different habitats, i.e., in lowland and highland habitats and in different types of land use systems, incorporating individuals that are subject to low and high exposure to sunlight, respectively. Investigate whether leaf and fruit production is affected by leaf harvest, by comparing the production rates of harvested and not harvested individuals.

5. Investigate *Krameria lappacea* in the field and examine anatomy and morphology of all root structures found. Document and compare the conditions of populations in locations subject to differently intense types of commercial extraction.

1.5.3 Overview

This dissertation consists of five manuscripts, which are either published in, submitted to or in preparation to be submitted to peer-reviewed scientific journals. Consequently, the Chapters 2[a] to 6[e] are structured as journal articles, each containing a separate introduction as well as sections for materials and methods, results, and discussion. All references cited in this thesis are given in a combined reference list after Chapter 7 and supplementary data for individual chapters are provided in the Appendix of this work.

The Chapters 2[a] to 4[c] are dedicated to studies on the economic botany of palms (Arecaceae) in northwestern South America (i.e., Bolivia, Colombia, Ecuador, and Peru), which are presented in chronological order, according to the date of publication and submission, respectively. Chapter 2[a] represents a standardized research protocol, which was designed and used for the obtention of data on harvest, production, and trade of palm raw materials and products by means of interviews with stakeholders that are involved in these areas of activity. Chapter 3[b] is a review on trade in native palm raw materials and products in northwestern South America, which is based on information from scientific literature, official statistics, and data that were acquired by means of interviews performed in Peru (in 2009) and in Bolivia (in 2010) using the standardized research protocol presented in Chapter 2[a]. Chapter 4[c] is a case study on productivity and management of *Phytelephas aequatorialis*, which presents biological data from several field trips to western Ecuador conducted by Danish and Ecuadorian colleagues (between 1991–1993 and 2011–2012, respectively) as well as basic economic data acquired by the author through interviews performed across western Ecuador (in 2011) using the standardized research protocol presented in Chapter 2[a].

---

[a] Brokamp G., Mittelbach M., Valderrama N., Weigend, M. 2010. Gathering data on production and commercialization of palm products. *Ecología en Bolivia* 45(3): 69–84. ISSN 1605-2528. Homepage: http://www.scielo.org.bo/scielo.php?script=sci_serial&pid=1605-2528&lng=en

[b] Brokamp G., Valderrama N., Mittelbach M., Grandez-R. C.A., Barfod A.S., Weigend, M. 2011. Trade in Palm Products in Northwestern South America. *The Botanical Review* 77(4): 571–606. http://dx.doi.org/10.1007/s12229-011-9087-7

[c] Brokamp G., Borgtoft Pedersen H., Montúfar R., Jacome J., Weigend M., Balslev H. subm. Productivity and management of *Phytelephas aequatorialis* (Arecaceae) in Ecuador. Submitted to *Annals of Applied Biology* on 11.07.2013, accepted with minor revisions on 12.07.2013, final revision ahead, AAB-2013-0230.

The chapters 5[d] and 6[e] are dedicated to basic botanical research on *Krameria lappacea*, a wild harvested and endangered medicinal plant from the Andean deserts in South America. Chapters 5[d] and 6[e] are presented in chronological order, according to the date of publication and submission, respectively. Chapter 5[d] is focused on the hemiparasitc nature and haustorium structure of *K. lappacea*. Here included are data on host range from field studies conducted in Andean sites across Peru (years 2004–2012) and results from anatomical (microtomy and light microscopy) and micromorphological studies (scanning electron microscopy) on roots and haustoria that were performed in Berlin. In Chapter 6[e] data on abundance and population structure from different locations are presented on the basis of field studies conducted in Andean sites across Peru (years 2004–2012). Additionally, Chapter 6[e] includes aspects of seed ecology resulting from field observations and a germination experiment that was carried out in a greenhouse in Berlin.

In Chapter 9 (Conclusions), all findings regarding the economic botany of native palms from northwestern South America and the results of conducted basic botanical and ecological research on *Krameria lappacea*, both already discussed in previous chapters, are summarized.

---

[d] Brokamp G., Dostert N., Cáceres-H. F., Weigend M. 2012. Parasitism and haustorium anatomy of *Krameria lappacea* (Dombey) Burdet & B.B. Simpson (Krameriaceae), an endangered medicinal plant. *Journal of Arid Environments* 83: 94–100. http://dx.doi.org/10.1016/j.jaridenv.2012.03.004

[e] Brokamp G., Schwarzer C., Dostert N., Cáceres-H. F., Weigend M. in prep. Now, where did all the Rhatanies go? Abundance, seed ecology, and regeneration of *Krameria lappacea* from the Peruvian Andes. To be submitted to *Journal of Arid Environments* or a similar journal.

2 Standardized data collection on trade in palm products

2.1 Introduction

This study is focussed on commercial aspects of Neotropical palms, palm products, product (pre-)processing and value chains; with the aim to characterize the current trade of palm products and their likely development in the future. One of the primary objectives was the design and development of a standardized research protocol (SRP; Appendix A) as a basis for the collection of significant and interoperable data on commercialization of palm products in the countries under study (Bolivia, Colombia, Ecuador, and Peru), across the different palm species and their products.

Palms provide a huge variety of products, ranging from construction material through domestic implements and fibre products to raw materials for food and cosmetics (Stagegaard *et al.*, 2002; Balslev *et al.*, 2008; see Chapter 3.3). The degree of commercialization is even more divergent than the products themselves and covers a range extending from the personal use and local barter to national and international trade, and from direct consumption to complex processing. As a result, total volumes harvested and marketed, and also the degree of sophistication in harvest, processing and commercialization vary widely. However, there are few detailed data on the absolute and relative economic importance of individual commodities and products, as well as on the corresponding value chains.

The main objectives in designing this protocol were the following: (I) to achieve a consensus between the desire for a maximum of useful data and the need for interview forms that can be applied in practice without losing time and without exhausting respondents; (II) make them widely applicable, regardless of product, species, role of the informant, region, and scope of marketing. After testing of the protocol in the field and a first check of the data obtained, all data from the different interviews were captured and summarized in a corresponding data capture table for subsequent interpretation and processing.

2.2 Protocol design

Already existing protocols for obtaining this type of data are – generally – focusing on a single value chain and need different types of forms for the various actors (e.g., Guel & Penn, 2009). Our protocol aims to be universal, i.e., applicable to any actor and product, so we only need a single universal form for each to collect relevant data at any level. The standardized research protocol (SRP) was designed to capture a representative set of data for each respondent, including: amount and source of raw material and type of product commercialized, value chain, trade routes, type of (pre-)processing, market and limiting factors of trade. This protocol was developed in six stages: (I) Identification of types of relevant data, (II) design of draft versions, (III) field testing of draft versions in Colombia and Peru (GB, MM and NV), (IV) revision and sending of protocols to the FP7-PALMS project partners for comments, (V) second revision and presentation in the FP7-PALMS project workshop (Villa Tunari, Bolivia), (VI) third revision and incorporation of changes suggested in Villa Tunari.

The draft protocol version turned out to be quite large and unwieldy when tested in the field, so we needed to remove and rearrange questions in order to reduce the overall number of questions, and thus the time needed per interview. The final version consists of only two forms of one page each so a personal interview can be done in 20–60 minutes, depending on product complexity and role of the respondent in the trade with palm products. This final version was also tested in field studies in Bolivia in early August 2010 (GB, MM). It was verified that this version is easy to use in the field and allows to obtain a maximum of relevant data in a minimum of time.

2.3 Structure and use of the protocol

In its final design, this protocol can be applied for structured interviews with open and semi-open questions. In this way, the protocol consists out of three components that are used in data collection.

2.3.1 Interview forms

The interview forms consist of a master sheet and an annex (Appendix A2), one page each. There is (only) one master sheet to be filled in for each interview, which compiles basic data (e.g., interviewer, interviewee, date, exact geographic location, role of interviewee in commerce of palm products, list of species/products). For each product a separate annex is filled in; the annex summarizes all the quantitative and qualitative data corresponding to each product, such as the amount of raw material that is used in absolute and relative terms (ratio of raw material and finished product), how and where raw material is harvested, which (pre-)processing steps are performed, what costs and benefits incur, where and to whom the raw material/product is sold and how it is transported, and what are the limiting factors of production/sale of a given product.

2.3.2 Manual

A manual provides detailled instructions on how to use the interview forms and contains a questionnaire that spells out all necessary questions (Appendix A1).

2.3.3 Data capture table

For further processing and interpretation, obtained interview data need to be tabulated. For this process an EXCEL spreadsheet was designed, i.e., a data capture table (DCT; Appendix A3) that corresponds to the interview forms, in order to facilitate raw data transfer without necessity of prior reorganisation or interpretation of data. Once completed, every line of the DCT corresponds to a distinct product (commonly more than one product is registered per interview), which simplifies the direct comparison of equal or highly similar products coming from different palm species, interviewees, and locations of the study.

2.4 Data storage and exchange

This SRP may be applied in order to obtain data by means of interviews. The original interview sheets completed during field work are archived and the data is transferred to the DCT, which permits an easy exchange among investigators and provides the basis for further processing and evaluation of the data. If participating researchers complete all their information correctly, the resulting data tables should be fully compatible and the unification of several data sets into a single database is just a matter of „copy and paste". Accordingly, the DCT is the only format that should be used for data transfer.

2.5 Problems and limitations

Despite supreme effort to ensure best usefulness of the here presented SRP, its implementation is characterized by inherent limits. Particularly, primary producers of raw materials and also representatives of small handicraft manufactures often only have a limited comprehension of amounts of time and resources that they invest in the different work processes of harvest, manufacture, transport, and sale. A standardization of investments per amount of commercialized product is here rather seldomly applied and thus exact data on time and money needed for each work process often remain unavailable. Time, that is invested in the production and sale of individual products, may vary extremely, and also deriving stakeholder income may differ drastically among months and years, which should be considered when performing an interview. Own observations should always be made, e.g., when interviewees perform relevant work processes, such as harvest or other types of management (Guel & Penn, 2009). Obtained interview data should be considered information from verbal reports with well known associated problems, such as lack of objectivity, poor memory, or imprecise articulation. Therefore it is recommended to verify obtained interview data with help of additional information sources (Yin, 2003), which may be literature, own observations, and a comparison with data from similar studies or data obtained from other stakeholders that participate in the same business.

In the here presented SRP some data (e.g., prices, amounts, cash income) are solicited intentionally as duplicative entries in overlapping questions to later be able to evaluate the data consistency. By that a rough estimation of the level of confidence for individual data entries is possible; an internal control, e.g., through a closer look on the answers to the following questions: „What is the total amount of the finished product that is sold per unit time?", „What is the sales price per unit finished product?", and „What is the total income from the finished product per unit time?" In theory, the sold amount of a finished product per unit time multiplied by the price should equal the revenue per unit time. In practice, however, there may be minor or major discrepancies between these figures reprted by individual stakeholders, which should be contolled, and data sets that were identified as incoherent should be discarded.

By contrast, in more elaborate business sectors (e.g., the trade with palmito or tagua, see Chapter 3.3.8 and Chapter 4.41), respondents of all stages of the production chain are generally well aware of exact figures on, e.g., trade volumes, costs, and benefits. However, often there is limited willingness to share that information with outsiders, as there is a fear that such information is misused by competitors; and some stakeholders simply consider data of that kind as confidential business secrets. Quantity and quality of confidable data obtained, thus, depends largely on the interviewer's demeanour, performance, and commitment, apart from the interviewee's knowledge and motivation, and scores from the data tests on consistency, as mentioned above.

Overall, it is crucial to overcome the reluctance of stakeholders to share information, which may be accomplished through a clear explanation of the motivation and objectives of the study in order to build trust on the part of participating interviewees. It is important to ensure that respondents understand that the interviewer's desire neither is to steal business secrets, nor to establish restrictive regulations, or to affect the trade in palm products. The goal is to obtain basic data, which is needed to be able to give recommendations and assistance in the medium-term development of a healthy and sustainable market. Currently, many wild palm species are exploited beyond sustainable limits. In order to ensure the continued existence of the market for palm products and – if possible – to foster its medium-term growth, a sustainable use of raw material sources is of vital importance.

2.6 Perspective

Data on production and commercialization of palm products obtained from small industries and commerce represent a meaningful basis for the description of the current situation and the possible future development in the trade with palm products, and will be crucial for studies on sustainability and formulation of policies: In NW South America, the current wild harvest of palms is not sustainable and while harvested amounts and the relative and absolute importance of exploited species largely remain unknown, it is very difficult to identify the most relevant target species and thus formulate realistic strategies for conservation and research in order to achieve sustainable commercial exploitation. Furthermore, while there are numerous laws that regulate and limit the extraction of wild resources in the four countries under study (Bolivia, Colombia, Ecuador, and Peru), there is only limited knowledge and understanding of the actual effect of these laws. The here presented protocol was designed with the purpose to partially fill this gap in our understanding.

Trade volumes and the relative and absolute importance of palm products are the most important arguments when confronting the administration and politicians with matters of research, conservation, and sustainability policies. The more extensive and cohesive data on production and marketing of palms are available, the more persuasive they will be to trigger economic and political changes.

3 Trade in Palm Products in North-Western South America

3.1 Introduction

Conservation through use or through trade has been proposed as a key mechanism to provide incentives for the conservation of species and habitats by turning them into sources of income (Peters et al., 1989; Wild & Mutebi, 1996). Sustainable harvest and trade of Brazil nuts and rubber are examples of this. Reconciliation between conservation and rural development can also be achieved via trade of genetic resources under access-and-benefit sharing agreements ensuring back-flow of cash generated by the trade. The most direct approach to conservation through use is sustainable trade of natural resources such as medicinal and aromatic plants (MAPs), wild fruit, and fibres. Palms provide both palm wood and a wide range of non-timber forest products (NTFPs). Many useful palm species occur in large, dense stands and have large regenerative potential. In this way they constitute ecologically and economically important natural resources that can be traded and may improve the livelihoods for rural populations.

Palms provide many useful products and literally thousands of individual palm uses have been reported in the scientific literature (see Balslev & Barfod, 1987; Bates, 1988; Balick & Beck 1990; Bernal, 1992; Henderson, 1995; Moraes-R. et al., 1995; Johnson & the IUCN/SSC Palm Specialist Group, 1996; Borchsenius et al., 1998; Macía, 2004; De la Torre et al., 2008; Soler-Alarcón & Luna-Peixoto, 2008; Galeano & Bernal, 2010; Macía et al., 2011; see also Table 3.1). These uses vary greatly in overall economic importance and trade levels. Most species and many rawmaterials are used locally by ethnic groups and bartered outside the cash economy, if traded at all. Other products are traded on a minor scale locally or regionally, or on a wider, national scale. The most common use categories of traded palm products are food (fruit, palm heart, vegetable oil; see Fig. 3.1), construction material (timber, thatch; see Fig. 3.2), raw material for handicrafts (mainly fibres and seeds; see Fig. 3.3) and medicine (Borchsenius & Moraes-R., 2006; Sosnowska & Balslev, 2009).

A thorough understanding of value chains for palm products is crucial for the development of current and future markets. There is much literature on palm use in tropical America that provides insights into the socio-economic impact of palm products. Non-timber forest products, including many palm products, are accepted as important sources of income for rural dwellers, but quantitative information on the role that NTFPs play in local economies is virtually non-existent (Padoch, 1987; Pinedo-Vasquez et al., 1990). There are few studies on current trade volumes, the economic potential and the value chains.

The aim of this review is to provide an insight into the volume of palm trade in north-western South America, encompassing Colombia, Ecuador, Peru and Bolivia, and to assess its impact at different economic levels. We also wish to clarify how far current extraction of palm resources agrees with, or is amenable to the "conservation through use" principle. We will concentrate on the commercially most important and most intensively exploited native palms across the major use categories in Bolivia, Ecuador, Peru and Colombia. These are *Iriartea deltoidea* (timber), *Astrocaryum* spp. (fibre, fruit), *Euterpe* spp. (palm hearts, fruit; Fig. 3.1 G–I), *Mauritia flexuosa* (fruit, oil; Fig. 3.1 A, B), *Oenocarpus bataua* (fruit, oil; Fig. 3.3 G), *Lepidocaryum tenue* (thatch; Fig. 3.2 A, C, D, F–H), *Ceroxylon* spp. (religious ornaments), and *Phytelephas* spp. (vegetable ivory). Table 3.1 summarizes the most important species and uses of South American palms as treated in this review.

3.2 Materials and Methods

We reviewed more than 200 publications on tropical American palm uses and trade, and searched relevant websites (see literature list). Internet sites are the only up-to-date sources on many currently traded palm products and their prices and were therefore extensively consulted, in spite of their ephemeral nature and consequent disadvantages as sources. Special attention was paid to trade with palm products in Bolivia, Colombia, Ecuador, and Peru. The referenced information was organized into a number of Excel®-spreadsheets. A core spreadsheet contained all product and species-specific information. Quantitative data were converted into the metric system. Values and prices were calculated in US$ (exchange rate as of November 2010). Scientific palm species names follow Govaerts & Dransfield (2005).

3.3 Results

3.3.1 Timber

Many tropical American palms have woody stems with hard and durable timber that is used for floors, walls, roof beams, boards, furniture, fishing and hunting tools, fruit boxes, and fence posts (Fig. 3.2 B, C, G–I).

Iriartea deltoidea, probably the most important palm species for timber in northwestern South America, is common throughout our study area (Henderson, 1995). It is mostly old palm trees that are cut since the quality of the timber increases with age, due to increasing amounts of sclerified tissues (Borgtoft Pedersen & Skov, 2001). *Iriartea* timber is extremely hard, durable and heavy, and mainly used for flooring and walls, and to a lesser extent for (fence) posts, roof beams and furniture (Borchsenius *et al.*, 1998; Borgtoft Pedersen & Skov, 2001; Moraes-R., 2004; Balslev *et al.*, 2008). It is also used to make tools for cultivation, hunting and fishing, banana props and fruit boxes (Barfod & Balslev 1988; Barfod & Kvist

1996; Anderson, 2004). *Iriartea* timber is increasingly used for handicrafts and furniture (Anderson, 1998) and is sold on local and national markets in Colombia (Galeano & Bernal, 1987) and exported to the United States, especially from Ecuador and Colombia.

Table 3.1 Focus species and their primary uses as treated in this review

Species	Primary use(-s)	Secondary uses	Trade level
Iriartea deltoidea (also *Socratea exorrhiza*)	timber (construction, furniture)	fruit (food, beverage & fodder), leaves (thatch), seed (handicraft)	local, regional, international
Astrocaryum chambira, A. malybo, A. standleyanum, A. murumuru, A. jauari	leaf fibre (handicraft)	fruit (oil, food, beverage & fodder), press cake (fodder)	local, regional, international
Mauritia flexuosa	fruit (food, beverage, oil)	fibre & seed (handicraft), timber (construction)	local, regional
Oenocarpus bataua	fruit (food, beverage, oil)	rhachis (construction, handicraft)	local, regional, international
Lepidocaryum tenue	leaves (thatch)	-	local, regional
Euterpe precatoria, E. oleracea	palm heart (food)	fruit (food, beverage), seed (handicraft)	local, regional, international
Ceroxylon spp.	leaves (ceremonial)	timber (construction)	local, regional
Phytelephas aequatorialis, P. tenuicaulis, P. macrocarpa	seed (handicraft)	leaves (thatch), fruit (food, beverage, fodder)	regional, international
Bactris gasipaes	palm heart & fruit (food, beverage)	-	local, regional

Fig. 3.1 Food from palms

Food from palms. (A–C) *Mauritia flexuosa*. a, habit (Iquitos region, Peru). (B) sale of *aguaje* fruits at a fruit market in Lima (Peru). (C) stem wound. (D) *suri*, the edible larva of *Rhynchophorus palmarum*. (E–I) *Euterpe* spp. products. e, preparation of *pepiado* near Isquandé (Colombia). (F) *açaí* soft drink (Iquitos, Peru). (G–I) *palmito*. (G) worker collecting *palmito* (Colombia). (H) river mole as *palmito* collecting point (Colombia). (I) canned *palmito* in a Peruvian shop.

Fig. 3.2 Use of palm products in construction

(A) *Lepidocaryum tenue* leaf piles for *crisneja* production. (B) *Socratea exorrhiza* stem splits (*ripas*) for *crisneja* production. (C) *crisneja* plaiting. (D) stockpiled *crisnejas*, sales unit *el ciento* (100). (E) roof ridge made of fronds of *Attalea* spp. (F) *crisneja* transport from producing community to Iquitos (Peru). (G) shop for construction materials selling *crisnejas* in Iquitos. (H) palm roof in an indigenous community near Iquitos. (I) palm roof made from *Geonoma deversa* leaves, Parque Nacional de Carrasco (Bolivia).

43

Fig. 3.3 Handicraft, pharmaceutical and cosmetical preparations from palms

(A–F) handicraft. (A) *Astrocaryum chambira* fibre extracted from young leaf shoot. (B) freshly dyed fibre. (C) cord of twisted fibre. (D) bracelets made with *chambira* fiber, *Phytelephas* spp. endosperm or *Euterpe* spp. seeds. (Peru). (E) vase woven from *Astrocaryum standleyanum* fibre (Colombia). (F) mats made from *A. standleyanum*. (G–I) cosmetics. (G) *Oenocarpus bataua* mesocarp oil (*aceite de majo*, Bolivia) as hair tonic. (H) *Attalea speciosa* (*cusi*) and *Attalea phalerata* (*motacú*) oils as hair tonic. (I) soap produced from *Attalea speciosa* oil.

In Ecuador a 10 m stem of *Iriartea* sold for 10 US$ in 1996 (Anderson & Putz, 2002; Anderson, 2004). A skilled worker is able to harvest 20 stems per day, so palm timber harvest represents a good daily income, compared to the average daily pay for unskilled labour, which is less than 10 US$ (Anderson, 2004). Little is known about retail prices of *Iriartea* timber and furniture in South America, but in the United States *Iriartea*-products are quite expensive. An office desk sells for ca. 1,000 US$ and kegs for 18 US$ each (Anderson & Putz, 2002). The price for raw materials makes up only 2–3% of the price of the finished product. Even when costs of transport, labour and additional materials are taken into consideration, the retail price is high. The primary producer receives only a modest share of the profits generated and the raw material trade is strongly influenced by local limitations in infrastructure. In Ecuador *Iriartea* has been depleted in several areas, not least since it requires an estimated 100 years to reach harvestable size and also because regeneration is poor in pastures and fallows, where it is typically harvested (Wollenberg & Inglés, 1998). Efficient policies for sustainable harvest and reforestation are not in place and the perspectives of *Iriartea* as a source of income for local and regional economies are bleak. A "fair trade" arrangement, by which the primary producers are guaranteed an adequate proportion of the final price, could accelerate the process of depletion. A maintenance and possible development of the national and international markets for *Iriartea* timber requires explicit and rigid harvest and reforestation strategies, based on reliable sustainability studies.

Ceroxylon timber is used for construction in Colombia (Albán *et al.*, 2008; Galeano & Bernal, 2010), but is not yet subject of large-scale export even if it is highly appreciated regionally. The trade with *Ceroxylon* timber is becoming highly lucrative because over-exploitation is leading to rising prices. Where natural stands are almost depleted, a cultivated palm stem fetches up to 50 US$, as compared to 10 US$/stem in areas where natural populations are still abundant (Pintaud & Anthelme, 2008). In Colombia, the stems of *Ceroxylon* quindiuense were formerly exploited as an important source of wax, however, this practice is now rare (Madriñan & Schultes, 1995) and we have no information about trade volumes or prices for this activity.

3.3.2 Thatch

All over tropical South America rural houses are thatched with palm leaves (Fig. 3.2) from *Euterpe precatoria, Geonoma deversa, G. orbignyana, G. macrostachys, Iriartea deltoidea, Oenocarpus bataua, Phytelephas macrocarpa, Attalea butyracea* and many more (Balslev & Barfod, 1987; Henderson, 1995; Borchsenius *et al.*, 1998; Flores & Ashton, 2000; Moraes-R., 2004; Borchsenius & Moraes-R., 2006).

Lepidocaryum tenue (*irapay*) is probably the most important species used for palm thatch. It is a small, rhizomatous, clonal palm occurring in lowland forest on terra firme, or on periodically inundated flood plains (Henderson, 1995; Scariot, 2001). *Irapay* is of particular importance in the greater Iquitos region of Peru, where tens of thousands of houses are thatched with its leaves. Although *irapay* leaves for thatching are of moderate economic importance overall, they constitute the highest ranking NTFP in many communities in terms of percentage of total households marketing (31%) and contribute considerably to the local economy, sometimes representing the most important source of cash income (Pyhälä et al., 2006). Locally, it may be the only NTFP contributing to household cash incomes (Mejía, 1992). Due to its clonal reproduction it may occur at high densities and may even dominate terrace palm communities (Vormisto et al., 2004; Balslev et al., 2010a). *Irapay* has distinct local and regional markets, but does not reach national and international markets. Shingles consisting of a slat with leaves attached to it are used throughout the Amazon and are referred to as *crisnejas* (Fig. 3.2 A–C; Mejía, 1983, 1988, 1992; Mejía & Kahn, 1996). Overall figures for the trade in *crisnejas* and its regional economic impact are not available so far, but it is evident that the local and regional socio-economic importance of *crisnejas* far surpasses its importance at a national level. For the roof of a 35 m2 house 20,800 leaves are required. In Peru, *crisnejas* are produced for local trade and sold at ca. 26 US$/*el ciento* (100 *crisnejas*; Fig. 3.2 D) or it is transported to major towns (Fig. 3.2 F), e.g., Iquitos and sold for ca. 45 US$/*el ciento* (prices for *crisnejas* with a length of 3 m with 40–60 leaves each, Brokamp et al., 2010a; 20 US$ for *crisnejas* of 2.2–2.5 m, Mejía & Kahn, 1996; see also Kahn & Mejía, 1987; Mejía, 1992; Fig. 3.2 G). A detailed study by Warren (2008) reported slightly different figures: Primary producers in the Iquitos region manufactured an average of 20–30 *crisnejas* per day and used 90–130 leaves for each *crisneja*. They earned 9–70 US$/*el ciento* (= 1.80–21.00 US$ per day). Vendors in Iquitos sold an average of 2,955 *crisnejas*/month with a profit range of 5–32 US$/*el ciento* and the consumers paid 23–120 US$/*el ciento*. In December 2009, primary producers of the Iquitos region used only approximately half the number of leaves per 3 m-*crisneja* (40–60) for the *crisnejas* sold, but still used over 100 leaves on *crisnejas* for their own houses (Brokamp, personal observation). The number of leaves used per (commercialized) *crisneja* (of 3 m in length) dropped from an average of ca. 100 (Warren, 2008) to an average of 50 (Brokamp et al., 2010a). With our limited data it is unclear if the lower number represents a drop in the quality of commercialized *irapay*-thatch in the last years in the Iquitos region. If there is a reduction in quality, this may be indicative of incipient resource depletion or go back to other market forces.

The harvest impact of *Lepidocaryum* leaves is considerable. Individual plants produce on average less than two new leaves/a, of which only one can be harvested without damaging the plant (Navarro, 2009; Navarro et al., 2011). Both sexual and clonal reproductive potentials of *irapay* are low, but population growth rates are greater than or not significantly different from 1.0, indicating populations maintained or increased in size in spite of the intensive harvest (Warren, 2008). Current levels of *irapay* harvest appear sustainable, but more detailed long-term studies would be required to test this assumption (Warren, 2008). The mid-term prospects for trade with this resource thus clearly depend strongly on the establishment and maintenance of sustainable harvest strategies.

3.3.3 Leaves Used for Ceremonial Purposes

The yellow spear leaves or young unfolding leaves of several *Ceroxylon* species are harvested to produce traditional, religious ornaments for processions on Palm Sunday (Borchsenius *et al.*, 1998; Moraes-R., 2004; Pintaud & Anthelme, 2008; Galeano & Bernal, 2010; Montúfar, 2010). Harvest, processing and sale of *Ceroxylon* leaves is an attractive, albeit highly seasonal, business. Individual leaves fetch ca. 0.5 US$ in the field and 24 US$ when processed into ornaments (Montúfar, 2010). Detailed data on the value chains or the extent of the trade in *Ceroxylon* leaves are not available.

3.3.4 Fibre

Numerous palms provide strong and durable fibres that are used for many purposes such as fishing nets, brooms and brushes, hammocks, carpets, bags and baskets, jewelry cases, adornments, and hats (Fig. 3.3 C–F). Fibre producing palms include *Leopoldinia piassaba*, *Aphandra natalia*, *Attalea colenda*, *Mauritia flexuosa* and several species of *Astrocaryum* (Balslev & Barfod, 1987; Henderson, 1995; Borchsenius *et al.*, 1998; Kronborg *et al.*, 2008; Guel & Penn, 2009; Isaza-A. *et al.*, 2010). Historically, fibres of *Leopoldinia piassaba* were traded to Europe and were the economically most important source of palm fibres (Spruce, 1860). Nowadays, *Leopoldinia* fibres are of local importance only (Putz, 1979; Bernal, 1992; Lescure *et al.*, 1992). In Europe palm fibres have been replaced by either plastic or, to a smaller extent, other natural fibres from annual crops such as hemp, linen and millet. The international trade in handicrafts, which includes mats and baskets made from palm fibres, exceeds 1 billion US$/a (www.intracen.org, accessed 20.11.2010). However, the product categories are rarely broken down and up-to-date export figures for palm-based handicrafts are not available for any of the countries in our study region.

Astrocaryum chambira is common and produces leaf fibres used in handicraft production in Amazonian Peru, Colombia and Ecuador (Borgtoft Pedersen, 1994; Holm Jensen & Balslev, 1995; Borgtoft Pederson & Skov, 2001; Gupta, 2006; Albán *et al.*, 2008; Guel & Penn, 2009; Isaza-A. *et al.*, 2010). *Astrocaryum malybo* and *A. standleyanum* are important sources of fibre in the Pacific lowlands of Colombia, Ecuador and Panamá (Borgtoft Pedersen, 1994; Velásquez Runk, 2001; Linares *et al.*, 2008; García *et al.*, 2010). *Astrocaryum* fibres are mainly used as raw material in cottage industries for handicrafts contributing considerably to local and regional incomes. In Colombia the annual export of handicraft amounts to ca. 400,000 US$ and includes numerous palm products (www.intracen.org, accessed 20.11.2010). Ecuador had a considerable export of *Astrocaryum standleyanum* fibres and processed items made from these fibres in the 1980ies (1981–1991, mainly to Peru and Japan) reaching maximum volumes of 37 t/a (1987), corresponding to an export value of 80,000 US$/a (Borgtoft Pedersen, 1994).

In Colombia, handicrafts made primarily of fibres extracted from A. malybo and A. standleyanum are sold at local markets or at arts and craft fairs in the major cities (Torres-R., 2007; Barrera-Z. *et al.*, 2008). Retail prices are two to three times higher than the producer's prices and these do not adequately reflect the artisans investment of time and raw material (Torres-R., 2007; García *et al.*, 2010). In Ecuador, where *A. standleyanum* is commonly left as a shade tree in agroforestry systems, the potential annual harvest of young leaves ranged from 579 to 1158 kg/ha/a (with two leaves harvested/palm/a) and 1,158–4,060 kg/ha/a (with four leaves harvested/palm/a), corresponding to 82–289 US$/ha/a respectively 165–577 US$/ha/a (based on market prices in April 1992; Borgtoft Pedersen, 1994). Both raw material and manufactured fibre-products are commercialized: young leaflets, from which fibres are extracted, are sold at 1.26–1.48 US$/kg, large hammocks are commercialized for 15 US$/kg (at 3.7 kg equals ca. 4 US$/kg). It is reported that an entire family (two adults, four children), preparing fibres and making hats as their main occupation, earned as little as 18 US$ per week without deducting the expenses for production material. Conversely, the production costs of the landowner employing harvesters at the minimum wage, is only about 15% of the income he obtains from selling the fibres (Borgtoft Pedersen, 1994). In Colombia, there is a growing concern about the sustainability of fibre extraction from *Astrocaryum* species, which is an important source of income for households in the region. Several initiatives with the purpose of encouraging sustainable fibre harvest and enrichment planting are under way (Penn & Neise, 2004; Barrera-Z. *et al.*, 2008; Guel & Penn, 2009; Torres-R. & Avendano-R., 2009).

In Ecuador and Peru, *Astrocaryum chambira* fibres are extracted from young leaves and processed manually. Harvest, fibre processing and handicraft production are often done by those who sell the finished products directly to the consumer. After separation of fibres they are bleached in hot water, washed and left in the sun for drying and further bleaching over 1–2 days (Fig. 3.3 A) (Bianchi, 1982; Paymal & Sosa, 1993; Holm Jensen & Balslev, 1995; Coomes, 2004; Linares *et al.*, 2008; Guel & Penn, 2009; Brokamp *et al.*, 2010a). Dying with natural or artificial dyes (Fig. 3.3 B) and twisting of fibres into thread (Fig. 3.3 C) requires an additional day. For a single hammock 1.8 kg fibre is required, which takes three, 8-hour working days to prepare. Lack of raw material and lack of time are the primary limiting factors for production of *chambira* based handicrafts. Therefore basic mechanization could undoubtly increase the volume and economical impact of the entire industry. In coastal Ecuador, for comparison, simple machinery has made processing of *A. standleyanum* fibre dramatically more efficient (Borgtoft Pedersen, 1994; Holm Jensen & Balslev, 1995).

The overall production time for a single *chambira* hammock is 5–8 days (Vormisto, 2002; Coomes, 2004) and it sells for 10–75 US$ in retail shops. The highest prices are fetched in towns with more tourist visitors (Holm Jensen & Balslev, 1995; Castaño *et al.*, 2007), but most handicrafts are bartered to river traders or merchants in the towns in exchange for daily goods. The trade is poorly organized due to the remoteness of the processing sites, much to the disadvantage of the primary producers. In Peru, the producer obtains the highest price when handicrafts are sold directly to tourists (hammock 9.5–30 US$, bag 1–5 US$), much less when selling to an intermediary in Iquitos (hammock 7.6–9.5 US$, bag 0.8–3.0 US$), and least when selling or bartering to river traders (hammock 5.0–7.6 US$, bags 0.8–1.9 US$) who visit the producing communities (Vormisto, 2002; Coomes, 2004). The situation is similar in Ecuador, where retailers buy carrying bags for 1.5–5.0 US$ and sell them for 2.5–10 US$, and buy hammocks for 12.5–15.0 US$ and sell them for 20–50 US$ (Holm Jensen & Balslev, 1995). In some parts of Peru and Ecuador the sale of *chambira* products represents a monthly income of (0–)82(−275) US$/household or 300–400 US$/household/a (Coomes, 1996; Coomes & Barham, 1997; Brokamp *et al.*, 2010a), which constitutes a large proportion of the overall cash income (Coomes, 2004). The sale of *chambira* handicrafts is the most important source of cash income in many lowland communities in Colombia, Peru and Ecuador (Bennett *et al.*, 1992; Borgtoft Pedersen & Balslev, 1992; Vormisto, 2002; Castaño *et al.*, 2007). The local use and sale of hammocks from *Astrocaryum chambira*, however, is limited, and cotton hammocks are often preferred, even in communities that produce *chambira* hammocks (Vormisto, 2002).

Further development of the *Astrocaryum* fibre market depends on the availability of raw material. Traditionally, entire palms were cut down to harvest the fibres, which has led to a severe decrease in population density in many areas. The most heavily exploited species are becoming increasingly rare in some areas due to destructive harvesting practices (Velásquez Runk, 2001; Coomes, 2004; Torres-R., 2007; García *et al.*, 2010). Fortunately, there are also several reports of simple nondestructive harvest methods (e.g., Borgtoft Pedersen, 1994) in some regions (Holm Jensen & Balslev, 1995; Torres-R., 2007; García *et al.*, 2010). It is increasingly being realized by local communities and public authorities that *Astrocaryum* fibres may be a finite resource, unless the species is sustainably managed. As a consequence, several initiatives have been started to resolve management issues and associated problems, such as land tenure (Guel & Penn, 2009). As in the case of *Lepidocaryum* thatch, the national economic importance of *Astrocaryum* fibres may be limited, but it still makes up one of the most important sources of cash income for many households in rural Colombia, Ecuador and Peru. The local and regional socioeconomic importance of *Astrocaryum* palms is therefore considerable (Bodmer et al., 1997). In some areas, artwork and handicraft are driving local economies and the demand for *chambira* fibres is steadily growing (Guel & Penn, 2009). Furthermore, there is a growing market for *chambira* fibres abroad, especially in France and

Germany. Clothes are now made of a mixture of *alpaca* wool and *chambira* fibres (www. ponchisimo.com, 10.09.2010). The future prospects for this unique fabric and other novel application of *chambira* fibre are, at present, difficult to judge.

3.3.5 Palm Heart

Palm heart, palm cabbage or *palmito* is a specialty vegetable, obtained from several palms (Fig. 3.1 G–I). It is extracted from the crownshaft, formed by the overlapping, tubular leaf sheaths, and consists of the immature, etiolated leaves. The nutritional value of palm heart is low, but it is a good source of dietary fibre (Mora-Urpí *et al.*, 1997). Palm heart is one of the economically most important non-timber forest products exported from north-western South America and the single most important edible palm product from native palms in this region. The volume of the world trade in palmito was 132.6 Mio US$ in 2008, with annual growth rates of 16% during 2004–2008 (Anonymous, 2009). Originally, single-stemmed *Euterpe edulis* from Brazil was the most important species delivering *palmito*, probably followed by *Prestoea acuminata* and *Euterpe oleracea* from Ecuador. *Euterpe edulis* is now rare and commercially extinct due to overharvesting (Kahn & Henderson, 1999; Backes & Irgang, 2004) and *Prestoea acuminata* has also suffered severely (Borgtoft Pedersen & Balslev, 1990, 1993). *Palmito* currently entering international markets is harvested from wild populations of *Euterpe precatoria* (mainly in Peru and Bolivia) and *E. oleracea* (mainly in Colombia and Brazil). The market share of palm heart extracted from plantation grown *Bactris gasipaes* is, however, growing and mainly so in Ecuador and Costa Rica. Already in the second year of production, *B. gasipaes* orchards yield ca. 1.35 t/ha/a *palmito* (Mora-Urpí *et al.*, 1997). Palm heart is traded at all economic levels. France is the main port of entry into Europe. *Euterpe precatoria* has been considered as one of the economically most important native species in Peru due to the high sales prices fetched for palm heart (Stagegaard *et al.*, 2002).

Until recently, the bulk of traded palm heart was obtained by destruction of wild palm stands. After the depletion of *E. edulis* in Brazil, *E. precatoria*, another singlestemmed species, largely replaced it in trade. Multi-stemmed *E. oleracea* is also exploited, but to a smaller extent. Since *E. oleracea* is able to regenerate after cutting, it is theoretically amenable to the development of sustainable management techniques (Vallejo *et al.*, 2010, 2011). Recently, *palmito* from cultivated *Bactris gasipaes* is replacing palm heart from *Euterpe* species harvested from the wild. This has happened both on the domestic markets in Colombia and the export markets, where Ecuador and Costa Rica are the main players. Palm heart production based on wild *Euterpe oleracea* is still considerable in Colombia, mainly for export.

Because of its economic importance, palm heart is probably the best understood palm resource in South America, although some studies fail to distinguish between palm heart obtained from the different species. In both Bolivia and Peru a massive industry was built up in the 1990ies for canning palm hearts that were extracted mostly from the native *E. precatoria* (Mejía, 1992; Stoian, 2004; Vormisto, 2002). In 1991, Peru exported 677 tonnes of canned palm hearts (Fig. 3.1 I) valued at over 1.5 Mio US$ (Anonymous, 2000). At this era palm heart was considered a product of great national importance (Pyhälä *et al.*, 2006). Based on data obtained from a single canning factory in Iquitos (interview by C. A. Grandez R.) trade peaked during the years 1996–2000, with a production of ca. 1,000 t/a and a value of up to 3.8 Mio US $/a. Subsequently, the annual production of palm heart decreased to 142 t/a in 2002 with a value below 300,000 US$/a. Since then there have been signs of slow recovery. If we assume the average weight of the individual palm hearts to be 500 g, the production figures of a single canning company in a peak year corresponded to 2 Mio felled palms/a.

In Colombia, export of palm heart is essentially based on *Euterpe oleracea*, which is harvested from the wild, whereas palm heart for domestic consumption is either extracted from cultivated *Bactris gasipaes* (Janer, 2002a, 2002b) or imported from Ecuador. Exact trade figures are not available, but export volumes for *E. oleracea* apparently peaked in the 1980ies, when nine canning factories processed 80,000 stems per day, which corresponds to nearly 30 Mio palms/a at a value of >4 Mio US $. Production dropped dramatically in the 1990ies (Vallejo *et al.*, 2010). Between 2000 and 2009 exports ranged from less than 3 t/a at a value of < 7,000 US$ (2003) to more than 500 t/a at a value of nearly 1.5 Mio US$ (2007; www. proexport.com.co, accessed 20.11.2010). This corresponds to a total of ca. 5 Mio felled palms/a based on an estimated average weight of 100 g per *E. oleracea* palm heart. Since *E. oleracea* is a multistemmed palm it produces new suckers from the base, unlike *E. precatoria*, which is single-stemmed and does not recover from the harvest. Detailed studies of the value chain in Colombia are not available. In Colombia the average harvest rate per person is 150 stems/day, and the price fetched by the harvester is 0.1 US$/stem, and the average monthly income for a harvester working 3–4 days a week is 122–162 US$, corresponding to approx. half the minimum wage of 262 US$/month (Vallejo *et al.*, 2010). Primary producers thus earn little from harvesting palm heart in Colombia, a situation that will persist as long as there are no alternative sources of income for the population groups concerned.

Palm heart trade is particularly well studied in Bolivia (Anonymous, 2010a). Until 1993 extraction rates were moderate (500,000 palm hearts/a), but increased dramatically to 7.3 million palm hearts/a in 1997 (Stoian, 2000) and subsequently dropped to less than 1.5 million palm hearts/a in 2004 (Stoian, 2004). Export of palm heart from Bolivia reached a maximum of 12 Mio US$ in 1997 and 1998 (Anonymous, 1999), making it the second most important non-timber forest product after Brazil nuts. In 2008, Bolivia exported a total of 3,580 tonnes of palm heart worth 9.4 Mio US$ (Anonymous, 2010a). Most Bolivian palm heart was harvested from wild stands of *Euterpe precatoria*. In any given area typically 90% of the mature trees were felled during the harvest period (Zuidema & Boot, 2000; Stoian, 2004). The minimum age of the trees felled for palm heart was estimated at 70 years and the average age at 90 years (Peña-Claros, 1996; Zuidema, 2000). Recovery of natural stands is slow. Under favourable conditions full population recovery is achieved in 75–80 years (Zuidema & Boot, 2000). Cultivated stands of *E. precatoria* are ready for harvest after 5–6 years (Villachica, 1997) or 12 years (Kahn & de Granville, 1992; Stoian, 2000b), which puts the destruction of wild populations into perspective as particularly irresponsible. Uncontrolled harvest will inevitably lead to depletion of natural stands (Kvist & Nebel, 2001). Palm heart extraction for trade in Bolivia is highly lucrative for local communities, in spite of the fact that the collector receives only 2–6% and intermediaries 3.4% of the retail price of the final product (Stoian, 2004). The break-down of the income derived from the export of canned palm heart from Bolivia to Brazil is as follows: retailers (40–52%), wholesalers (13–36%), exporters (14–21%), and canning plants (8.3–16.9%; Stoian, 2004). Nevertheless, 21% of the benefits generated by export of palm heart are returned to rural areas in northern Bolivia (Kahn, 1988). Destructive harvest of palm heart from *Euterpe precatoria* has led to large-scale destruction of the natural stands in Bolivia and Peru (Peña-Claros, 1996; Moraes-R., 1998), and probably also in Ecuador and Colombia. The collapse of the Bolivian canning industry was, however, largely due to poor sanitary conditions in the canning plants and external economical factors such as currency crises in importing countries (Stoian, 2004). Ecuador used to export large quantities of wild harvested palm heart from *P. acuminata* and *E. oleracea* (ca. 900 tonnes in 1991) at a commercial value of 1.5 Mio US\$. Today, wild harvested palm heart only plays a minor role (Borchsenius & Moraes-R., 2006) and Ecuador's palm heart trade is mainly based on cultivated *Bactris gasipaes* (Anonymous, 2000, 2009; Anonymous, 2010a). The volume of the palm heart trade in Ecuador has increased steadily since 1997 at which time it was worth 12 Mio US\$/a (Anonymous, 2000). In 2008, the export alone reached a value of 72.7 Mio US\$/a. Ecuador is now the largest exporter providing 55% of the canned palm heart in international trade, followed by Costa Rica with 20% (Anonymous, 2009). Palm heart harvested from the wild is loosing importance on international markets. In Colombia wild harvested palm heart provided <1% of the internationally traded palm heart in 2008. The corresponding figures for Peru and Bolivia are less than 3.5% and 7.5%, (Anonymous, 2010a). Unsustainable wild harvest of palm heart may soon lose its socio-economic importance.

3.3.6 Fruit

Several palm fruits are part of the staple diet of rural populations in north-western South America. Palm fruits are rich in starch, high-quality protein and oil, but low in acids and sugar and have high nutritional value (Balick & Gershoff, 1981; Balick, 1985, 1986, 1992; Bora *et al.*, 2001; Miranda *et al.*, 2008; Jacobo *et al.*, 2009; Oboh, 2009). They are consumed unprocessed, boiled, blended with water, or used as fodder for domestic animals. Palm fruits are traded locally, regionally and, to a lesser extent, nationally.

Bactris gasipaes (peach palm). The most important palm fruit in tropical America, *B. gasipaes*, is widely cultivated and doubtfully known from the wild (Clement & Arkcoll, 1985; Clement & Mora-Urpí, 1987; Blanco-Metzler *et al.*, 1992; Mora-Urpí *et al.*, 1997; Borgtoft Pedersen & Skov, 2001; Couvreur *et al.*, 2007; Balslev *et al.*, 2008). The yield may reach 20–30 t/ha/a in well managed stands (Mora-Urpí, 1979). Numerous land races differ in fibre, oil, and carotene contents, and fruit size (Henderson, 2000; Clement *et al.*, 2004; Jatunov *et al.*, 2010). *Bactris* fruits have a high content of carbohydrates (60–80% dry weight; Mora-Urpí *et al.*, 1997) and the starchy fruits are boiled and consumed directly in large quantities (Borchsenius & Moraes-R., 2006). Numerous attempts have been made to process the fruits further into products such as flour and fodder due to their high carbohydrate content. Some of these products have already entered the market (Mora-Urpí *et al.*, 1997; De Oliveira *et al.*, 2006). However, in spite of domestication and intensified research on *Bactris gasipaes*, the market has developed little in the past decades (Clement *et al.*, 2004).

Mauritia flexuosa. Wild populations of *M. flexuosa* (*aguaje* or *buriti*) are an important source of edible palm fruit in north-western South America, next only to cultivated *Bactris gasipaes* (Peters *et al.*, 1989; Penn, 2008). *Mauritia flexuosa* has a wide distribution in north-western South America and is also very common in Brazil. It often forms extensive, monodominant stands, called *aguajales*, in periodically inundated areas. The total area of *aguajales* in Peru alone is estimated at 5.3 million ha of which five million are in the Department of Loreto. In natural habitats the trees reach the fruiting stage in about 8 years and maintain a high productivity for 30–40 years, after which productivity declines. A mature female tree produces ca. 290 kg of fruit/a (Anonymous, 2005). In Roca Fuerte, Peru, the annual harvest of *aguaje* fruits is 1 t/ha (Macuyama-R., 2008) and the productivity of wild *aguajales* in Colombia may be 9.1 t/ha/a and that of plantations 19 t/ha/a (Castaño et al., 2007). Fruiting of *Mauritia flexuosa* (Fig. 3.1 A) is aseasonal, with a peak that differs among localities (Navarro, 2006). *Aguaje* forms part of the staple diet in lowland Peru and Brazil with large turn-overs of fruits at local and regional markets, whereas the markets in Bolivia, Ecuador and Colombia are comparatively small (Castaño *et al.*, 2007; Holm *et al.*, 2008). The export market is negligible constituting less than 1% of the production in Peru. Traditionally, *aguaje* fruits form an important part of the diet of Amerindian groups (Delgado *et al.*, 2007). The trade has recently expanded to markets of major towns and *aguaje* is increasingly being sold in the capital Lima. The main market is in Iquitos, however, where fruits are sold in various degrees of processing: crude, cooked, as *aguaje* soft drink called *aguajina*, or in fermented form, ice creams, popsicles (*chupetes*), or frozen in plastic bags (Kahn, 1991; Mejía, 1992; Del Castillo *et al.*, 2006; Navarro, 2006; Delgado *et al.*, 2007). *Aguaje* flavoured ice cream is a major product, mainly sold in the Iquitos region, but also in Pucallpa and Lima (Rojas-R. *et al.*, 2001). Harvesting of *aguaje* fruits represented the third most important economic activity for households in Roca Fuerte, Peru, in 2002, accounting for 31% of the cash income and involving 75% of the households (Manzi & Coomes, 2009).

Aguaje fruits (Fig. 3.1 B) are sold in Iquitos in bags of 35–40 kg. Several fruit varieties are recognized on the market. *Aguaje shambo* with a thick, red mesocarp is the superior quality. Another variety is *ponguete*, with a more yellowish pulp. Varieties with particularly thick mesocarp are collectively referred to as *aguaje carnoso* (Ruiz-M., 1991). Additional descriptive terms such as *aguaje de color* and *posheco* are also used. Prices vary according to season. During peak harvest time (July–October) they are sold for 0.06–0.07 US$/kg and in the low season for 1.5–1.7 US$/kg (Anonymous, 2005). This price difference is partially reflected in retail prices for the *masa de aguaje* (pulp), which is used for ice cream. The price/bag (600–700 g) varies between 0.5 and 1.2 US$/kg (Rojas-R. *et al.*, 2001; Anonymous,

2005) over one season, depending on the supply and demand. These prices are similar to those reported from Colombia (ca. 0.5 US$/kg; Castaño *et al.*, 2007) and Bolivia (0.28 US$/kg, personal observation G. Brokamp and M. Mittelbach). Prices also depend on fruit maturity and fruit quality.

Harvesters typically sell aguaje fruits to riverboat traders in their home villages, or wholesalers in Iquitos (Macuyama-R., 2008) and in Peru there is a complex market pattern for both processed pulp and unprocessed *aguaje* fruits (Anonymous, 2005). For each type of finished product the value chain included both primary collectors, several levels of intermediaries, wholesalers, street vendors and retailers. Intermediaries (*mayoristas*) in Iquitos may subcontract patrones in local communties, who in turn subcontract collectors. The harvest is sent by boat from the collection sites to markets in the vicinity of the Iquitos river ports. There are also intermediaries in Iquitos (*rematistas*) buying *aguaje* directly off the boats and barges. Sometimes they even use speedboats to meet the *aguaje* boats before they reach Iquitos and purchase the fruit with cash. The amounts bought typically range between 10 and 30 bags per person and the sellers are typically independent collectors without a business network in Iquitos. The *aguaje* is resold, either to other intermediaries, or directly to the consumer, usually through family members. A number of small enterprises have specialized in selling fruits, *masa de aguaje* (i.e., pulp), or processed products such as ice cream. A major part of the fruit entering the markets are cooked in street kitchens and sold as a snack (Del Castillo et al., 2006).

Detailed data for sale and consumption are not available and estimates vary, since informal street vending makes up a major proportion of the *aguaje* commerce. It has been suggested that approximately 30 tonnes of *aguaje* are consumed daily in Iquitos, roughly equivalent to the fruit obtained from just over 100 trees. This translates into a per-capita consumption of 2.14 kg/month (Anonymous, 2005) and an estimated annual overall consumption of 10,000 tonnes corresponding to the annual yield from ca. 38,000 trees. Other estimates suggest a consumption of 150–660 tonnes per month (5–22 tonnes per day; García & Pinto, 2002; Delgado *et al.*, 2007) and an annual trade for all of Peru of 10,000 tonnes, of which only a tiny fraction is exported (<1 t/a; Santa Natura, www.aguajeperuano.blogspot.com, accessed 05.09.2010). Seasonal fluctuations in prices, differences in *aguaje* varieties as well as details of the value chain leading to the finished product are not fully understood. A conservative estimate indicates that the raw *aguaje* trade in Iquitos alone is worth around 550,000 US$/a (based on lowest price of 2 US$ and an average weight of 37.5 kg/bag). However, it may be several times higher – reaching 2.5 Mio US$/a under the assumption of an average price of 9.5 US$/bag and an average weight of 37.5 kg/bag. The overall economic impact including processing, transport and retail will likely be several times higher and may reach several million US$/a.

Destructive, large-scale harvesting is depleting this seemingly inexhaustable natural resource. Every year 24,000–200,000 palms are cut down with the sole purpose of harvesting the fruits. *Mauritia* is dioecious and since only the mature, female trees are cut down for their fruit, the proportion of male and juvenile trees is steadily increasing in the accessible *aguajales*. The market is contracting in many regions of Peru because readily accessible *aguajales* have largely disappeared and only a small number of productive female plants remain for fruit extraction (Guel & Penn, 2009). Prices in Iquitos are rising due to an increasing demand in combination with a reduced supply. Fruit quality is highly variable and the best varieties such as *shambo* are cut down first (Manzi & Coomes, 2009) and they are becoming increasingly rare in the wild and on the market. Furthermore, the natural *aguaje* stands are genetically impoverished since only trees producing low-grade fruits are left to reproduce. Large-scale degradation of *aguajales* and concomitant genetic erosion pose immediate threats to future attempts to commercially develop this promising resource. Nondestructive harvesting techniques must be introduced to stop this direct and indirect resource depletion. Considering the popularity of *aguaje* fruits, semi-domestication, domestication and cultivation in agroforestry systems (Penn & Neise, 2004) will probably be the only options for meeting the current and future demand. *Aguaje* orchards would not compete with other crops, since *Mauritia* grows on marginal lands, such as swamps, that have little agricultural potential.

Euterpe spp. *Açaí*, the fruits of *E. precatoria* and *E. oleracea*, has developed from a minor local product into an international commodity in the past 10 years. In southwestern Colombia, fruits of these *Euterpe* species are highly appreciated and sold for ca. 0.36 US$/kg in March 2010 (Vallejo *et al.*, 2010). The market potential for *Euterpe precatoria* fruits is promising. This species produces 13–20 kg fruits/plant/a (Bovi & de Castro, 1993), which represents a commercial value several times higher than that of its palm heart. The fruits of *Euterpe precatoria* are traded locally in Amazonian Colombia at a price of 0.54 US$/kg (Castaño *et al.*, 2007). In Bolivia fruits for oil extraction are sold by street vendors for 0.36 US$/kg. The corresponding price on the retail market is 0.57 US$/kg. For local collectors this is a lucrative business, since they are able to harvest 50 kg in 3 h (Madre Tierra de Amazonia/IPHAE, pers. comm.), corresponding to an income of ca. 6 US$/h. Individual palms of *E. oleracea* may produce nearly 27 kg fruit/a. This is equivalent to a market value of 9.8 US$/palm/a (on the local market), which is considerably more lucrative than the once-only income of 0.1 US$/stem obtained by felling the tree to obtain the palm heart (Vallejo *et al.*, 2010). When *palmito* production based on *Euterpe oleracea* was at its peak in Colombia in the 1990'ies, a single year's palm heart harvest removed fruits corresponding to an approximate market value of 290 Mio US$. Although this amount of fruit likely would have exceeded the local market capacity at the time, they could have formed the basis for an export business. In Brazil, the *açaí* export became important over a decade ago and individual companies now produce up to 1,000 t/a (www.acai-mania.com, accessed 20.11.2010) and *açaí* fruits constitute the single most important food item (by weight) in certain communities in Brazil (Murrieta *et al.*, 1999). Fruit harvest is, however, seasonal and can not provide a continuous income to rural populations, in contrast to the aseasonal harvest of palm heart. In pulp production only about 30% of the *Euterpe* fruit is retrieved. The pulp is sold both locally and regionally, for example to *açaí* ice cream factories at about 2 US$/kg in Bolivia (Madre Tierra de Amazonia/IPHAE, pers. comm.).

Recently, *açaí* pulp was introduced to the Asian, European, and US markets, as an ingredient in "energy drinks" (www.calidris28.com, accessed 12.11.2010), fruit juices and yogurt (Coïsson *et al.*, 2005; Sabbe *et al.* 2009a, 2009b). It is widely traded in the US and gaining market shares in Europe (e.g., in Germany) with considerable market potential (Sabbe *et al.*, 2009a, 2009b). Managing *açaí* is highly profitable in Brazil (Muñiz-Miret *et al.*, 1996). For one hectare of orchard the profit ranged from 896 to 1,814 US$/a, after deducting leasing expenses. The açaí market is currently dominated by E. oleracea from Brazil (Vallejo et al., 2010) with an estimated production of at least 480,000 t/a (Rogez, 2000; Brondízio, 2008). The estimated volume of the international market of *açaí* pulp was approx. 30,000 metric tons in 2007 (www.biocomerciosostenible.com/Boletin2.html, accessed 17.10.2010), and rising fast. One website (www.alibaba.com, accessed 20.11.2010) lists 38 wholesalers of *açaí* pulp, of which 24 are based in Brazil and five in India. Wholesale prices vary from 6 to 270 US$/kg and individual companies can deliver up to 2,000 t/a. Retail prices for fruit powder in Europe and North America vary from 70 US$/kg for large orders to 100–340 US$/kg for smaller orders. Promotion of *açaí* products is boosted by alleged high contents of phenolic compounds with beneficial "antioxidant" properties (Schauss *et al.*, 2006; Pacheco-Palencia *et al.*, 2008), although health authorities such as EFSA (Anonymous, 2010b) regard these claims with considerable scepticism and recent studies revealed that various açaí drinks are only slightly higher in antioxidants than, e.g., apple juice, and considerably lower than other and cheaper fruit juices (Seeram *et al.*, 2008). The current boom may therefore be short-lived, but it is difficult to predict how the overall *açaí* market will develop in a longer time perspective.

The largest and most accessible stands of *E. precatoria* and *E. oleracea* have been mostly destroyed for *palmito* harvest. The palms are also still commonly cut down for fruit harvest in rural areas throughout north-western South America. This practice terminally precludes ecologically and economically sustainable development of this valuable resource (Velarde & Moraes-R., 2008). A sustainable development of markets for wild harvested *açaí* depends on the introduction of sustainable harvest policies. Also, the perishable fruit needs speedy transport to processing plants, and overall a considerable degree of technical sophistication and development will be required to establish a major *açaí* industry in the Andean countries.

The prospects for palm fruits as a commodity in the food industry depend on a variety of factors such as availability of raw materials, formation of prices, harvest practices, and sustainability of resource management. The high tocol and carotenoid levels may also provide an opening to the lucrative health food and functional food markets on the international level.

3.3.7 Palm Oil from Native Species

There is much interest in oil from palm pulp and kernels (Pesce, 1985; Lleras & Coradin, 1988; Bereau *et al.*, 2001, 2003; Jacobo *et al.*, 2009; Montúfar *et al.*, 2010). Although oil contents are often low, palm oil production per area is potentially high for native palms, reaching 0.5 t/ha (Lleras & Coradin, 1988). Most native palm oils have a fairly conventional fatty acid composition, with chain lengths of (6–)16–18(–24) of mainly saturated fatty acids. Many palm kernel oils are rich in lauric acid (C12), and, based on their fatty acid composition, most fall within the lauric and myristic acid subclass and the palmitic acid subclass, which are primarily of interest as texture agents added to cosmetics and food (Dubois *et al.*, 2007). The fatty acid composition of most tropical American palm oils is similar to many known and widely used oils, but some palm oils are very high in tocols and carotenoids. Due to these particular properties, palm oils may be subject to further market development, provided that they are critically evaluated for negative side effects and that technologies for extraction and processing are refined. The market potential of native palm oils is currently severely limited by their generally high price (Table 3.2).

Attalea phalerata, *A. speciosa*, *A. maripa*, and *A. butyracea* are locally exploited in Bolivia to produce oils that are primarily used in cosmetic preparations at small and medium scale, but they could also be used in human consumption (Borchsenius & Moraes-R., 2006). In Bolivia, fruits of *Attalea speciosa* (*cusi*) for commercial processing are currently sold at 0.05 US$/kg and the oil is sold for about 30 US$/kg in bulk sale and 29 US$/liter in retail sale. *Cusi* oil is primarily used in soap, cosmetics and shampoo (www.indelcusi.com, accessed 20.11.2010; Fig. 3.3 H, I). The oil is marketed professionally as hair care product (www.oleunsbeauty.com, accessed 20.11.2010). Due to a particularly high percentage of lauric acid the demand for these oils for the use in cosmetics may increase in the future.

Mauritia mesocarp oil has high β-carotene, α-tocopherol, and oleic acid contents and resists oxidation, which is highly appreciated by the food and cosmetic industries (Lleras & Coradin, 1988; Santos, 2005; Vásquez-Ocmín *et al.*, 2010). Oil from the seed has a high concentration of ώ6 (linoleic acid), which is known to prevent negative effects of oxidation (Vásquez-Ocmín *et al.*, 2010). In Bolivia, an experimental plant for palm oil extraction uses 30 kg of *Mauritia flexuosa* fruits to produce 1 l of *buriti* oil (Madre Tierra de Amazonia/IP-HAE, pers. com.). Prices are high due to these meagre yields. In Brazil, 1 l of *buriti* oil is currently sold for 23–26 US$/kg in wholesale and for 130–210 US$/kg on the European retail market (www.regenwaldladen.de, accessed 20.11.2010; www.seasonsskin.com, accessed 20.11.2010). Due to the high prices *buriti* oil is imported into the US and Europe mainly for use in skin and hair care products (www.thebodyshop.co.uk, accessed 20.10.2010; www. oleunsbeauty.com, accessed 20.11.2010).

Oenocarpus bataua is widespread in north-western South America and represents a major natural resource (Balick, 1985; Kahn, 1991; Moraes-R. *et al.*, 1995; Miller, 2002; Montúfar & Pintaud, 2006; Castaño *et al.*, 2007; Orihuela-Ardaya, 2009). Fruits are harvested from the wild, often by felling the trees (Vasquez & Gentry, 1989; Borgtoft Pederson & Skov, 2001; Miller, 2002; Stagegaard *et al.*, 2002). A large proportion of the fruit is consumed locally (ca. 40% in Bolivia, Departments of Beni & Pando), the remaining part is processed industrially for oil (Fig. 3.3 G) or ice cream (Orihuela-Ardaya, 2009). Near densely populated areas the abundance of *Oenocarpus bataua* is rapidly decreasing. As a consequence fruits are rare on local markets and prices are rising (Vasquez & Gentry, 1989; Miller, 2002; Gupta, 2006). The fruiting cycle of *Oenocarpus bataua* is biennial and fruit set is relatively low. The yields in natural *Oenocarpus* stands in Peru and Ecuador range from 0.7 to 1.3 t/ha/a, which corresponds to ca. 50–100 kg of oil/ha (Kahn, 1991; Miller, 2002). On the Pacific coast of Colombia fruit yields were ca. 0.23 t/ha/a (Castaño et al., 2007). In Amazonian Colombia there is a considerable market for fresh fruits, which are sold at 0.54–1.6 US$/kg (Castaño *et al.*, 2007). The fruits are sometimes preserved and used for jams, ice creams and soft drinks (Vasquez & Gentry, 1989; Borgtoft Pederson & Skov, 2001; Albán *et al.*, 2008; Balslev *et al.*, 2008). Avaluable oil is extracted from the mesocarp with a yield of 6.5–12% of the fruit fresh weight. It is reminiscent of olive oil in fatty acid composition and contains high levels of tocopherols (Montúfar et al., 2010). *Oenocarpus* oil is consumed by humans and used in cosmetic preparations (Gupta, 2006).

The use of *Oenocarpus* oil is not new. A considerable export market existed in the early 20th century, when 100–200 t/a where exported from Brazil and Colombia to the USA and Europe. However, this market collapsed due to changes in market structure and destructive harvest (Balick & Gershoff, 1981). There is currently an enormous interest in developing the market for *Oenocarpus* oil, and the commercial potential is generally considered as high, even for regional and national markets (Mejía, 1992). In Peru, it is regarded as one of the

Table 3.2 Current prices and availability of palm oils

Some palm oils as raw materials for the cosmetics market (researched on the websites indicated, 10.11.2010)

Species, type	www.100amazonia.com (FOB, wholesale, US$/kg) BRAZIL	www.100amazonia.com (minimum order/-supply limit/year) BRAZIL	www.oca-brazil.com (retail, US$/kg) BRAZIL	www.camdengrey.com (wholesale and retail, US$/kg) US
Palm kernel oil (*Elaeis guineensis*)	-	-	-	3.7—6.6
Palm oil, bulk (*Elaeis guineensis*)	-	-	-	3.7—6.6
Coconut oil fractionated (*Cocos nucifera*)	-	-	-	7.5—13.1
"Babassu/Babaçú" oil (*Attalea speciosa* as *Orbignya oleifera*)	-	-	166	8.0—17.4
"Tucuma" seed butter/oil (*Astrocaryum aculeatum* as *A. tucuma*)	7.50—9.00	200/20,000	85	29.7—48.5
"Açai" butter/oil (*Euterpe oleracea*)	80.00—120.00	50/50,000	400	34.9—52.0
"Murumuru" butter (*Astrocaryum murumuru*)	12.95—16.69	200/50,000	85	45.7—83.8
"Buriti" Oil (*Mauritia flexuosa*)	21.88—23.75	200/50,000	166	-

most promising palms from a commercial point of view (Kahn, 1988, 1991). In 1996, the unrefined oil sold for 10–11.66 US$/liter and the purified oil for 32.50–40 US$/liter in Ecuador (Miller, 2002). The net value of the total amount of unrefined oil that can be obtained from a one hectare grove in 1 year varies between 500 US$ and 1,166 US$ and for the refined oil it may be as high as 4,000 US$ (in 1996 prices). The total market value of the *Oenocarpus* fruits themselves was estimated to be 115.92 US$/ha/a (Pyhälä *et al.*, 2006). A study of one hectare of forest at Mishana, RNAM, Peru recognized Oenocarpus fruits as the single most important non-timber forest product, accounting for 35% of the household income (Pyhälä *et al.*, 2006).

Cosmetics containing *Oenocarpus* oil are being marketed at a minor scale via the internet and on organic markets (*Bio ferias*) in Lima (www.mishkiperu.com, accessed 20.10.2010). In Bolivia, there are several projects dealing with industrial processing of *Oenocarpus bataua* fruits (Miranda *et al.*, 2008).

One of the areas with a well developed market for *Oenocarpus bataua* fruits is Riberalta in Bolivia. Collectors here sell 12–90 kg of fruits per day at a price of 0.26 US$/kg (Ortiz Camargo, 2007). According to different sources, 25–37 kg of fruits are required to obtain one liter of filtered oil (Madre Tierra de amazonia/IPHAE, pers. com.). The fruit required costs up to 9.5 US$/liter when purchased on the local market, and the overall production costs are estimated at 26 US$/liter (Miranda *et al.*, 2008). Retail prices for the oil in La Paz are around 78 US$/liter, which means that the costs of the raw material make up less than 12% of the retail price of the finished product. *Oenocarpus* oil is a lucrative product for the national market, even when additional costs for packaging and transport are considered. In Colombia, the retail price of the oil is 1–1.25 US$/kg and thus much lower (Castaño *et al.*, 2007). Ecuador has already engaged in exports of *Oenocarpus* oil to Europe, where it is used for cosmetic purposes (www.eza.cc, accessed 20.11.2010), mainly in haircare products (www.aromandina.com, accessed 10.11.2010; www.rahua.com, accessed 20.11.2010; www. eza.cc, accessed 20.11.2010; www.ainy.fr, accessed 20.11.2010).

The cake that remains after oil extraction of palms is highly nutritious, with elevated levels of protein (nearly 20% in *Attalea speciosa*) and/or carbohydrates (nearly 85% in *Bactris gasipaes*). The cake is commonly traded locally as animal fodder. Animal nutrition is under-developed in many South American countries and programmes focussing on the dual use of palm fruits for the extraction of palm oil and processing of the press-cake into high-quality animal fodder could have high socio-economic impact.

Improvement of the industrial processes involved in the oil extraction may contribute to re-ducing production cost. The high vitamin contents of some native palm oils may also be of importance on the domestic markets, due to serious problems with malnutrition and vitamin deficiency especially in the poorer part of the population. Nevertheless, the prospects for a large scale export of native palm oils for human consumption are not promising, since it is unlikely that they will be competitive in price and quality, compared to numerous established vegetable oils. Current prices for native palm oils are not competitive on the international market for food and food ingredients (Table 3.2).

3.3.8 Vegetable Ivory

Vegetable ivory is the collective term for the hard endosperm of palms, which is rich in hemi-celluloses and oil. In South America this product is obtained from species of *Phytelephas*, especially: *P. macrocarpa* (Peru and Bolivia), *P. aequatorialis* (Ecuador) and *P. seemannii* (Colombia). Vegetable ivory or *tagua* has a long history in international trade. Exports started in the 19th century and were mainly directed to Europe (Acosta-Solís, 1948; Pérez-Arbeláez, 1956; Barfod, 1989; Barfod *et al.*, 1990; Borgtoft Pedersen & Balslev, 1990, 1993). *Tagua* was among the five most important export commodities from both Colombia and Ecuador in the second half of the 19th century (Acosta-Solís, 1948; Bernal, 1992). Vegetable ivory can be readily turned, carved, polished and stained (Hofmann, 1995) and used for handicrafts (Fig. 3.3 D) and buttons. In Colombia, *tagua* trade started to decline around 1920 and by 1935 the industry had virtually disappeared. In Ecuador, exports peaked around 1929 reaching 25,000 t/a, valued at over 1.2 Mio US$, equivalent to about 15 Mio US$ in present day prices (Borchsenius & Moraes-R., 2006). Similarly, 1,000–2,000 tonnes of *Phytelephas macrocarpa* were exported from Peru between 1920 and 1940, mainly for the manufacture of buttons (Anonymous, 2002). During and after World War II the international demand for *tagua* dropped dramatically, because it was replaced by plastics (Barfod, 1989). Lately, since the 1990ies, the *tagua* trade is increasing again (Borgtoft Pedersen & Balslev, 1990, 1993; Pülschen, 2000). In 1991 the overall value of the vegetable ivory products in Ecuador was estimated at 4.2 Mio US$ (Borgtoft Pedersen & Balslev, 1990, 1993) and *tagua* is now exported from Ecuador, Colombia and Peru, mainly for the button industry. Manufacture of *tagua* buttons is usually wasteful since only 5–7% of the dry weight of the palm endosperm is retrieved in the finished product. Traditional fabrication (late 19th century) of *tagua* buttons was less wasteful and 3 kg of raw material went into the production of 1 kg of buttons (Hofmann, 1995).

In 1995, Ecuadorian harvesters received 4.35–6.25 US$ for one quintal (ca. 50 kg) of *tagua* and derived 40% of their monthly income from this product alone (Velásquez Runk, 1998). *Phytelephas aequatorialis* is the most important palm in terms of availability and accessibility, sustainability of the harvest as well as present and future commercial value (Borgtoft Pedersen & Skov, 2001). In Peru, there is an incipient market for *tagua*, based on *P. macrocarpa*. Peru exported 1.45 tonnes in 2002 increasing to 5.73 tonnes in 2005. The dehusked endocarps containing the seed are bought from village communities at 0.1–0.2 US$/kg depending on size (Navarro, 2006) and they are resold for more than 12 US$/kg, which is 60–110 times the price paid to the primary producer. The retail price for dried, unprocessed nuts in Ecuador is roughly 4–7 US$/kg depending on size (www.nayanayon.com, accessed 20.11.2010). On the German retail market individual *tagua* nuts weighing 30 g are sold for 3.5 US$, which corresponds to 115 US$/kg (www.taguagalerie.de, accessed 20.11.2010) or 600 times the price paid to the primary producer. In Europe, buttons made from *tagua* are gaining market shares in designer clothing. Individual *tagua* buttons (slice of nut with 2 holes, otherwise unprocessed) are sold at retail prices of 0.9–2.5 US$ in Germany, more elaborate designs for 6.75 US$ or more. In Ecuador, simple beads are sold for up to 90 US$/kg, more elaborate designs are sold for up to 929 US$/kg. On the German retail market carved figurines and boxes may fetch over 4,000 US$/kg (www.taguagalerie.de, accessed 20.11.2010), which is more than 30,000 times the value of the raw material in Peru and Ecuador.

Phytelephas stands vary in density and productivity (Navarro, 2006). *Phytelephas aequato-rialis* produces 4 t/ha/a in Ecuador, *Phytelephas seemannii* groves produce 2.25–12 t/ha/a in Colombia. In a forest reserve of 394 ha near Iquitos, Peru, vast populations of *Phytelephas macrocarpa* produce 0.77–1.67 t/ha/a. A conservative estimate for the average productivity of the entire area is 1.22 t/ha/a (Navarro, 2006). Thus, the resource available to the local communities can, under ideal circumstances, generate more than 60,000 US$ worth of raw material every year, corresponding to a potential export value of ca. 6 Mio US$ (in 2005 prices). The seeds are collected from the ground after rodents have eaten the mesocarp which represents sustainable harvest. If harvested before full maturity the vegetable ivory will be of inferior quality and lacks the characteristic colour and lustre (Borgtoft Pedersen, pers. comm.).

3.4 Discussion

3.4.1 Trade

Colombia, Ecuador, Peru and Bolivia all have a considerable international trade in palm products, and although exact figures are lacking, it is clear, that the two main products are palm heart (*palmito*) and vegetable ivory (*tagua*). For each country export values of these commodities amount to several million US$. The fruit of *Mauritia flexuosa* is the third most important palm product in terms of market volume. This is surprising, since it has a distinctly regional value chain and the trade is concentrated in the greater Iquitos region of Peru. Other South American palm products are subject to inreasing international trade. Oils derived from other native palm species are gaining importance but have a much smaller share of the inter-national market. At present, Brazil is the key player on the rapidly growing market for *açaí* fruits and a key-player on the *aguaje* market, not least due to development of new products.

The remaining palm products are exported in much smaller amounts and trade is difficult to quantify, since they do not appear in export statistics or are included in wider categories, to-gether with non-palm products (e.g., handicraft). Furthermore, trade at the national, regional and local levels is not usually captured in official statistical surveys. The best example of a really well-developed and massive regional trade in palm products is the *Mauritia flexuosa* fruit market in Peru. A number of products derived from these fruits are traded in tremendous volumes throughout the year. The processing industry is diversified and value chains are of-ten complex. *Mauritia flexuosa*, in both its crude and processed forms, is traded within Peru, with a growing market in cities far away from Iquitos, such as Lima. *Mauritia flexuosa* is the economically most important palm species in Peru. A complex evaluation system for fruit quality for different uses is in place, but technical sophistication of the industry is low. This is particularly true for harvesting techniques, which remain destructive and have caused and continue to cause a severe decline of the most valued palm populations.

There are also budding regional and national markets for several native palm oils such as those derived from *Oeonocarpus bataua, Mauritia flexuosa*, and various species of *Euterpe, Attalea*, and *Astrocaryum*. Extraction techniques are still primitive and inefficient. Production costs and prices are high, reaching more than 20 US$ per kg, so these palm oils may not develop into major commodities at national or international markets. However, native palm oils are increasingly used in cosmetics, both within the countries of origin and abroad. National markets in "native" cosmetics are rapidly growing. Perhaps as a consequence of rising prices, some companies have been able to meet acceptable quality standards.

Palm fruits are important at local and regional levels as food. They have high contents of oil, starch and protein and elevated carotenoid and tocol levels. This composition of nutrients makes them suited for both human consumption (staple diet, functional foods and food additives) and animal fodder, since malnutrition still constitutes a serious problem in northwestern South America.

Palm thatch represents another flourishing regional market for palm products. Thatch is a locally important product that is sold outside the region of origin, generating considerable cash income for producers and middlemen in Iquitos. Whereas the value chain and technology are well understood, the overall volume of this trade remains poorly known. Like other palm products in the Amazonian region of north-western South America thatch often constitutes the most important cash income for individual small-holders or entire communities. Similarly, handicrafts made of palm fibres or palm fruits may play a minor role at the national level and a moderate role at the local and regional levels, but they are crucial for the livelihoods in the communities where they are harvested and processed. Thus, several palm products (*Astrocaryum chambira* fibre, *Oenocarpus bataua* and *Mauritia flexuosa* fruit, *Lepidocaryum* leaves, *Phytelephas* nuts, etc.) provide the only or most important sources of cash income (often >30% of the entire cash income) for numerous local communities and are crucial for local – and sometimes regional – economies. Trade volumes, as measured in metric tons and US$ and as captured in official statistics, therefore do not adequately reflect the socio-economic importance of these products and grossly underestimate their importance.

3.4.2 Value Chains

Value chains are heterogenous, depending on product type, market penetration and the number of middle men involved. Value chains may differ even within the same region and for the same product. The bulk of palm products is marketed directly, i.e., the family that harvests the product sells it directly to the consumer at the local market. If processing is involved (manufacture of handicraft, processing into oils, etc.), then this will typically be carried out by family members. The primary producer then receives the added value (gross benefit). For handicraft sold at the national level a relatively high proportion of the retail price percolates back to the primary producers (*Astrocaryum chambira* carrying bags, Ecuador: 50–60%; *Astrocaryum chambira* hammocks, Ecuador: 30–60%; thatch, Peru: 55%). The proportion of the price for the raw material for the domestic industry going back to the sourcing price for the raw material is still considerable (e.g., 12%, *Oenocarpus bataua* fruit for oil extraction, Bolivia). With increasing trade volumes and industrialization, the costs of raw material in relation to the retail price becomes increasingly reduced. For canned palm heart (Bolivia) and furniture from *Iriartea* timber (Ecuador) the corresponding figures are only 2–6%. In Peru, the price fetched by the primary producer of the unprocessed vegetable ivory nuts is only 1.6% of its export price (FOB) and 0.1% of the price for nuts sold on the German retail market. Most of the profit is thus typically generated further up in the value chain, and often abroad.

3.4.3 Sustainability

Colombia, Ecuador, Peru and Bolivia have all implemented legislation for prohibiting "nonsustainable" use of forest resources (De la Torre *et al.*, 2011). There is, however, little evidence that these laws have a noticeable effect on how resources are managed. Legislation is largely ineffective, because the bulk of the trade in palm raw materials is informal. One positive example, however, is the apparently effective legal protection of species of *Ceroxylon* against overharvesting of leaves for ceremonial purposes (Galeano & Bernal, 2010). Also, vegetable ivory is being marketed based on presumably sustainable management practices to make it more attractive to the increasingly large segment of ecologically conscientious consumers. In spite of the marginal percentage of the retail price being returned to the primary producers, there is a rising awareness of the value of this resource and the importance of protecting it against overharvesting. *Lepidocaryum* leaves for thatch are increasingly being harvested using sustainable management practices. Fibre extraction from *Astrocaryum* is also increasingly managed sustainably instead of destructively. However, contrary to the intentions of legislation, the bulk of the trade in native palm products in north-western South America is still largely based on destructive harvest of wild populations, even in cases where this is not necessary (Bernal *et al.*, 2011).

A rough estimate based on the scattered annual production figures suggests that a minimum of several tens of million palm trees are cut down annually and, for the main part, unnecessarily to obtain fruits or leaves. Rising prices and reduced availability of raw materials is already widespread in a range of products such as timber and *Oenocarpus* and *Mauritia* fruits. Much of the palm extraction is therefore based on shifting extraction practices. New areas are harvested every year and the resources are already depleted near villages and urban areas.

The palm heart industry provides a particularly well-researched example of nonsustainable resource exploitation. Currently, tens of millions of palms are destroyed every year to deliver a limited quantity of a specialty vegetable. In terms of biomass destroyed in relation to biomass used, palm heart harvest is inherently destructive and wasteful. It is difficult to envisage how sustainable wild harvest should be managed, without causing a massive increase in the price of the raw material. Sustainable wild harvest is likely not an economically viable alternative. Replacing wild palm heart with palm heart from plantations may be desirable from a conservation point of view, since it reduces the pressure on native stands. However, large-scale cultivation has negative side effects such as clearance of forest and loss of income for local farmers involved in the wild harvest. Furthermore most of the profits generated by palm heart plantations typically end up in the hands of large land owners and private companies.

Unless destructive harvest is replaced by non-destructive practices, which are available for almost all of the species, even the most common palms such as *Mauritia flexuosa, Oeonocarpus bataua* and *Euterpe* spp. will face commercial extinction. Since destructive harvest is often selective, the best-quality trees are lost from the gene pool. As a consequence, genetic resources rapidly degrade and reduce the options for future resource development. Figures concerning the relative abundance of destructive harvest as opposed to sustainable harvest for other major palm products are not available, but are urgently needed for adequate resource management. There is a growing understanding of the negative impacts of destructive harvest and progress has been made in several regions towards implementing more sustainable harvest techniques. Small enterprises, such as "Madre Tierra de Amazonia (supported by IPHAE)" and "Indelcusi" in Bolivia and "Mishki" in Peru make an effort to ensure the sustainability of the palm products that they sell. Behind these enterprises are often NGOs, such as the "Fundación de amigos de la naturaleza" (FAN), WWF in Bolivia, or the "Rainforest Conservation Fund" (RCF) in the Iquitos region in Peru.

3.4.4 Conservation through Use

Many palm products have developed markets at various economic levels. Products such as timber, fibre, fruit and vegetable ivory, could likely increase their market share under adequate management of the resources, the value chains and the market. High productivity per area and local abundance of palms delivering a range of raw materials make them ideally suited for "conservation through use". Overall, the development, establishment and control of sustainable harvest techniques are probably the most important requirements for positive mid-term developments. Possible threats against this scenario are bad governance, corruption, and uncertainty of land tenure. Serious impediments in both policy making, law enforcement, and research will have to be overcome to develop the full potential on an ecologically and economically sustainable basis. Also, the value chains will need to be adjusted, so as to ensure that primary producers receive an adequate share of the overall benefits. This has to go hand in hand with strict policing and the clarification of contentious issues such as land tenure and ownership of the resources - otherwise higher profits will be an incentive for the informal and destructive harvest.

3.5 Conclusions and Recommendations

A wealth of studies exists on trade of native palm species in north-western South America, here used as a collective term for Bolivia, Colombia, Ecuador and Peru. In this review we focused on palm products of major economic importance on the local, regional, national, and international markets, attempting to cover major use categories. It is surprising that details on trade volumes and value chains remain incompletely documented, considering the socio-economic impact of palms to the livelihoods of millions of rural dwellers in South America. North-western South American palms provide a range of products that potentially could sustain flourishing industries. Many palm species are suited for production of vegetable oils, food supplements (high vitamin contents) and animal fodder (high protein and carbo-hydrate contents) and several species deliver valuable timber for construction, tool making and handicrafts.

The future of markets for native palm products is difficult to predict. Palms that were recently considered of marginal importance now deliver international commodities such as açaí fruits. Vegetable ivory once played a major role on the world market, but was substituted by plastics. Recently, this unique material has been rediscovered and now forms the basis of a booming industry. Resource depletion may, however, present a serious future impediment. Already with present trade volumes wild stands of several palm species cannot sustainably satisfy the demand. Another challenge is the ability of the primary producers to deliver raw material in predictable quantities and uniform qualities. Large-scale production for food, food additives and animal fodder will require preselection and propagation of cultivars with desirable characteristics. There will be an increasing pressure to move from extractivism to agricultural production with increasing trade volumes, either via agroforestry or plantations. This is particularly true for smaller and more densely populated countries, such as Ecuador.

We foresee that an increasing demand for any given palm product will not lead to a propor-tional increase of the income and hence increased socio-economic benefits in rural areas. It should be noted that destructive harvest and value chains are intricately linked: the commeri-cal value of palm products inadequately reflects the true value of the raw material in terms of availability and sustainability of harvest. Unfortunately, the benefit obtained by the primary producers is often so small that they are not encouraged to engage in sustainable (and more laborious) management and harvesting techniques.

Although market developments for timber and fibre have not been fully researched, our research suggests that technological innovation is needed to develop the industry beyond handicrafts. As for palm products used for food and cosmetics the problems are equally clear. Since most palm fruits perish quickly, the time of transportation between the harvesting and proccessing sites should be minimized, which is often difficult in the Amazon region and represents a serious logistic problem. Extraction of native palm oils is still carried out in technically unsophisticated ways, with fluctuating yields and poor-quality oils as a result. There have been numerous attempts to develop state-of-the-art extraction techniques, but an overall coordination is missing. Fruit production is mostly seasonal, which means that processing facilities can only be used part of the year, which in turn makes it difficult to recoup investments. The obvious solution to this problem is making the manufacturing plants more flexible, so that they are able to process several kinds of palm fruits (and possibly other crops). A major development of national and international markets will remain impossible as long as production and transport costs are as high as they are at present. The rapidly growing market for açaí fruit somewhat contradicts this, but we believe that prices are currently inflated by overstated health claims. Concerted action should be taken to ensure stable supply and uniform quality of native palm products at all levels of commercialization. Growing markets can only be served by the formation of cooperatives or direct-sourcing partnerships. Fair-trading could be an efficient marketing instrument to assure that benefits are funneled back to the primary producers and harvest occurs on a sustainable basis.

Fierce competition within South America represents a major obstacle for the countries included in this review in their attempts to obtain market shares. Brazil has already taken the lead, and is a serious competitor regarding all major palm products exported from Peru, Bolivia, Ecuador and Colombia. The low prices for vegetable ivory on the world market are at least partly due to competition between different north-western South American countries. A similar situation will probably arise as new products are developed – and rapidly copied elsewhere.

4 Productivity and management of *Phytelephas aequatorialis* (Arecaceae) in Ecuador

4.1 Introduction

Tropical ecosystems harbor thousands of useful plant species that are harvested and used on a daily basis by rural communities and which play a major role in securing the livelihoods of numerous people (Balslev, 2011). Palms stand out as a plant group of particular importance and usefulness in north-western South America (Macía *et al.*, 2011); production and commercialization of palm-derived non-timber forest products (NTFPs) often represents the most important or only source of cash income for local communities and individual households (Brokamp *et al.*, 2011), apart from the direct use and consumption of palm resources for subsistence. Accordingly, the commercial exploitation of palms is often fundamental for the ability of rural communities to participate in the cash economy in such crucial areas such as health care, education, and technology. However, the legal and administrative framework regulating the extraction and trade of NTFPs is complex and sometimes contradictory (de la Torre *et al.*, 2011). This, together with the absence of policing, contributes to a widespread overharvesting and mismanagement of palm resources in the Andean countries (Bernal *et al.*, 2011).

Phytelephas aequatorialis Spruce is a dioecious multipurpose palm endemic to western Ecuador where it grows from sea level up to 1500 m a.s.l. on the western slopes of the Andes (Barfod, 1991a; Borchsenius et al., 1998). In western Ecuador most natural forests had been cleared or degraded already 20 years ago (>95%; Dodson & Gentry, 1991) and the remaining forests are highly fragmented (Sierra, 1999; Weigend et al., 2005). Phytelephas aequatorialis occurs naturally in tropical rain forests and in (semi-)deciduous forests and cloud forests and is particularly common in disturbed areas and along rivers, including periodically flooded areas, where it often forms dense populations – the so-called *taguales*. *Phytelephas aequatorialis* was often spared when forests were cleared for agriculture and pasture (Fig. 4.1 A) because of the commercial value of this palm (Borgtoft Pedersen & Skov, 2001). Therefore, *P. aequatorialis* is now mainly found as a component of anthropogenic land-use systems replacing the original forest, particularly in the provinces of Manabí and Esmeraldas where it is a common component of pastures and mixed agroforestry systems (Fig. 4.1 C), together with cacao (*Theobroma cacao* L.), coffee (*Coffea arabica* L., *C. canephora* Pierre ex A. Froehner) and many other crops (Velásquez Runk, 1998; Lozada *et al.*, 2007). The stands on the floodplains of the lower Río Santiago in Esmeraldas (Fig. 4.1 E) are long and narrow (40–50 m wide) and limited to the natural levee along the river. In such places *P. aequatoria-*

lis is often subject to heavy management resulting in low regeneration and many more adult females than males (Velásquez Runk, 1998). Since the populations of this species are decreasing in parts of its range it has been listed as "near threatened", i.e., according to IUCN criteria it is close to qualifying for or is likely to qualify for a threatened category in the near future (Montúfar & Pitman, 2003). The species does, however, also occur in protected areas, including the Machalilla National Park, and the nature reserves Mache-Chindul, Cotacachi-Cayapas and Buenaventura.

The palm is commercially exploited mainly for its hard endosperm, known as vegetable ivory (*marfil vegetal* or *tagua* in Spanish), but also for its leaves, that are used for thatch and are locally called *cade* (Borgtoft Pedersen, 1994). Vegetable ivory has a long history in international trade (Brokamp *et al.*, 2011). Exports started in the 19th century, peaked in 1929 and dropped dramatically during and after World War II, mainly because vegetable ivory was replaced by plastics (Barfod, 1989). Since the late 1980s vegetable ivory has become a significant source of income again (Velásquez Runk, 1998) and the price for fresh, unpeeled seeds increased notably in the last two decades, from 1.50 US$ in 1988 (Barfod *et al.*, 1990) up to 5–15 US$ per *quintal* in 2011 (*quintal* = 45.36 kg). Exports of vegetable ivory reached a value of 14 million US$ in 2011, mostly as *tagua* discs (*animelas*) for the manufacture of buttons (BCE, 2012). This makes vegetable ivory the second most important product from native palms in Ecuador, after palm hearts (Brokamp *et al.*, 2011). Besides the use of vegetable ivory for *animela* production there is an extensive handicraft industry of figurines, jewelry, buttons, etc. and *tagua* is also used as industrial abrasive (Gonzales *et al.*, 2012) and as natural component in latex paint (Salazar Villón, 2006). Overall, vegetable ivory is a raw material perfectly suited for the environmentally conscious or "green" consumer market (Velásquez Runk, 1998). Therefore demand for vegetable ivory will likely continue to grow in the future. Corresponding trade figures on commercialization of leaves are not available since this commercial activity takes place at a local scale not captured in official statistics (de la Torre *et al.*, 2011).

Phytelephas aequatorialis is typically associated with a rich fauna feeding on the fleshy mesocarp, but not on the seed itself. This fauna includes squirrels (*Sciurus aestuans* L.), agoutis (*Dasyprocta* spp.), deer (*Odocoileus virginianus* Zimmerman), opossums (*Marmosa* spp.), porcupines (*Coendou* spp.), and pacas (*Cuniculus paca* L.), some of which hoard seeds in their burrows and act as the species' short distance dispersers (Barfod, 1991b; Velásquez Runk, 1995). The mesocarp is used locally as feed for domestic animals, to attract game, in traps for fish and rodents, and for human consumption (Koziol & Borgtoft Pedersen, 1993; Macía *et al.*, 2011). It is also used for extraction of edible oil, but only to a minor extent (Montúfar & Brokamp, 2011). Seed predation can be serious, especially in the highlands, and includes infestation by termites (Velásquez Runk, 1995) and bruchid beetles (Borgtoft Pedersen, 1995).

A productivity analysis over one year failed to demonstrate any influence of environmental variables on leaf or infructescence productivity (Velásquez Runk, 1998). In this paper we present data on productivity derived from 365 tagged individuals in six localities ranging from the coastal lowlands to the Andean slopes at 1400 m a.s.l near the upper elevational limit of the species. This study specifically addresses the following questions: 1) Does leaf and infructescence production in *P. aequatorialis* vary under different ecological conditions such as altitude, exposure to sunlight and type of management? 2) How does leaf harvest affect fruit production? On the basis of data on these topics, we hope to be able to suggest sustainable management practices.

4.2 Materials and methods

4.2.1 Study area

Observation on commercial uses, marketing, management practices and distribution were carried out throughout coastal Ecuador. We tagged 365 individuals of *Phytelephas aequatorialis* at six localities with different types of management (forest, pasture, and agroforestry) in the provinces of Cotopaxi, Esmeraldas and Manabí (Table 4.1). Population structure, and production of leaves and fruits were assessed in three plots situated on the Andean slopes (plots I, III, V) and three plots in the lowlands (plots II, IV, VI) using plot V as control for plot III (comparing the impact of leaf harvest). In three plots (II, III, IV), both male and female palms were tagged; in the other plots only females were tagged. Plants for which the sex could not be determined were excluded. Plots II and IV are located in an area that provides vegetable ivory for the factories in Manta and Portoviejo and there is a notable commercialization of leaves for thatch. In plot II leaves are harvested annually, while in plot IV the population of *P. aequatorialis* lies within a private nature reserve and since 2003 and there has been only sporadic harvest of leaves and fruits. Voucher specimen: Borgtoft Pedersen 67303 (AAU).

4.2.2 Field data

Field data was collected in 1991–1993 and again in 2011–2012. Details on production and marketing of vegetable ivory and leaves for thatch were obtained through 50 interviews of local harvesters, farmers and stakeholders mainly in the province of Manabí (using the protocol of Brokamp *et al.*, 2010b).

In five of the six plots, all palm individuals were registered and classified as seedlings (small plants still connected to the seed or bearing undivided leaves), juveniles (acaulescent plants with divided leaves), sub-adults (immature, caulescent palms), or adults (reproductive, caulescent palms) (following Balslev *et al.*, 2010b). In plot V, acting as control for impact of leaf harvest on fruit production, only adult, female palms were registered.

Fig. 4.1 Habitats and habit of *Phytelephas aequatorialis*

(A) Pasture (plot III). *P. aequatorialis* often represents the only woody component in pastures. (B) The globular infructescences were marked with paint. (C) *P. aequatorialis* in a mixed agroforestry system. (D) Disintegrating infructescence (*mococha*) of *P. aequatorialis*. (E) Dense *P. aequatorialis* stand (*taguales*) on the floodplains of the lower Rio Santiago area in Esmeraldas (where the agroforestry plot is located). (F) Seeds of *P. aequatorialis* on the ground.

Immature infructescences with a diameter >20 cm (>4–6 months old) were marked by painting the surface of one fruit in each infructescence (Fig. 4.1 B). After one year new infructescences were painted with a different color. Leaves were marked with a string or ribbon placed around the spear leaf and when plots were revisited the number of new leaves was counted as number of expanded leaves encircled by the marking. In addition three younger but fully expanded leaves were marked with paint on the petioles and the following data were recorded: number of leaves, number of infructescences and inflorescences, and exposure to light (see Table 4.1).

Tagged palms in the three highland plots were checked five times during two years and in the lowland agroforestry plot only at the end of the observation period, after 1.3 years. Tagged plants in plots II and IV were checked every second month during one year counting the number of infructescences shed, the number of new infructescences, the number of new leaves, and the phenological state of each individual was noted.

Functional lifetime of expanded mature leaves was calculated as number of leaves in the crown divided by number of leaves produced per year. The average time needed for an infructescence to mature was calculated as mean number of tagged infructescences divided by mean number of shed (mature) infructescences per year. In addition, in plots II and IV female reproductive organs were classified into ten developmental stages (Table 4.5), for each of the stages 20 reproductive organs were marked and the time of transition between stages was registered. When mature infructescences were shed, the number of fruit scars on the rachis (Fig. 4.1 D) and the number of seeds per fruit were counted. In plots I, III, V, and VI the dry weight of seeds and endocarps was determined after drying for 30 hours at 60°C.

4.2.3 Statistical analysis

We used Statwiev SE+ Grafics (1988) from Brainpower, Inc., Calabasas, California for analyses. Testing for difference between samples was performed with the non-parametric Mann-Whitney U-test (Elliot, 1977). The non-parametric Kendall rank-order correlation coefficient (Siegel & Castellan, 1988) was used when testing for correlation. Actual values of p are given when possible, p=0.05 is the general level of significance used. P-values mentioned in the text refer to the Mann-Whitney U-test, if not stated otherwise. Non-parametric tests were used, since most sampling sizes did not permit a confidential test for normality.

Table 4.1 Study areas

descriptive data from six plots with marked individuals of *Phytelephas aequatorialis* in western Ecuador. Vegetation types according to Holdridge and collaborators (1971). Exposure to light was estimated visually on a scale from 1–5 (mean±SE), 1 indicating total shade and 5 total exposure to sun light. *aequatorialis* in western Ecuador. Vegetation types according to Holdridge and collaborators (1971). Exposure to light was estimated visually on a scale from 1–5 (mean±SE), 1 indicating total shade and 5 total exposure to sun light

Plot (region, observation period)	Location	Land-use and vegetation type	Exposure to light	#tagged individuals	Treatment
I Forest plot (highland, November 1991–November 1993)	Prov. Cotopaxi, on the West Andean slopes at 1400 m alt. Near Palo Quemado, 10 km from pasture/harvested plot (00°22'S, 78°55'W).	Disturbed old growth forest with palms in partial shade. Wet lower montane forest.	Low exposure (mean: 2.4±0.2)	19 female	No treatment
II Forest plot (lowland, November 2011–November 2012)	Prov. Manabí, Junín at 93 m alt. Near road entry to Balsa Tumbada (00°56'S, 80°14'W).	Secondary forest (after 10 years of recovery). *P. aequatorialis* population with native vegetation and remnants of unmanaged cash crops. Tropical dry forest.	Low exposure (mean: female: 2.6±0.1; male: 2.7±0.1)	70 female 98 male	Only sporadic harvest of leaves and fruits.
III Pasture plot (highland, November 1991–November 1993)	Prov. Cotopaxi, on the West Andean slopes at 1325 m alt. Near San Francisco de Las Pampas (00°26'S, 78°57'W).	Pasture with most individuals of *P. aequatorialis* growing in a stand with little grazing between palms. Wet lower montane forest.	High exposure (mean: female: 4.7±0.1; male: 4.7±0.1)	20 female 19 male	Leaves harvested for thatch from all tagged individuals (4.6 leaves were left intact in the course of harvest).

Table 4.1 Study areas (continued)

Descriptive data from six plots with marked individuals of *Phytelephas aequatorialis* in western Ecuador. Vegetation types according to Holdridge and collaborators (1971). Exposure to light was estimated visually on a scale from 1–5 (mean±SE), 1 indicating total shade and 5 total exposure to sun light. *aequatorialis* in western Ecuador. Vegetation types according to Holdridge and collaborators (1971). Exposure to light was estimated visually on a scale from 1–5 (mean±SE), 1 indicating total shade and 5 total exposure to sun light

Plot	Location	Vegetation type	Exposure	Sex	Treatment
IV Pasture plot (lowland, November 2011–November 2012)	Prov. Manabi, between Canuto and Calceta at 37 m alt. Near road entry to Mocoral (00°49'S, 80°08'W).	Pasture with scattered individuals of *P. aequatorialis*, some cash crops, and native trees. Flat land. Tropical dry forest.	High exposure (mean: female: 4.5±0.1; male: 4.6±0.1)	38 female 30 male	Leaves harvested for thatch from all tagged individuals every year. Last harvest in September 2010. Seeds collected.
V Pasture plot (highland, November 1991–November 1993)	Prov. Cotopaxi, on the West Andean slopes at 1400 m alt. Near Palo Quemado, 10 km from pasture/harvested plot (00°22'S, 78°55'W).	Pasture with scattered individuals of *P. aequatorialis*. Wet lower montane forest.	High exposure (mean: 4.5±0.2)	21 female	No treatment
VI Agroforestry plot (lowland, May 1992–November 1993)	Prov. Esmeraldas, at 5 m alt. On the banks of Rio Santiago, near Maldonado (01°04'S, 78°54'W).	Agroforestry system with cultivated and spontaneously growing species. Canopy formed by large trees in some parts and by *P. aequatorialis* in other parts. Moist tropical forest.	Mixed exposure (mean for 28 palms with exposure ≤3: 2.3±0.1; mean for 22 palms with exposure >3: 3.9±0.2)	50 female	No treatment

4.3 Results

4.3.1 Population density and sex ratio

The populations of *Phytelephas aequatorialis* varied substantially in density (total number of individuals per hectare), proportion of age-classes, and sex ratio (Table 4.2). The pasture plot (IV) had the lowest and the forest plot (II) had the highest density, largely due to an exceptionally high number of juveniles documented in the latter. Very similar and highest numbers of adult individuals were found in the forest (II), in the pasture (III), and in the agroforestry plot (VI); by far the lowest number of adults was documented in the forest (plot I). Number of subadults was highest in the agroforestry plot (VI), while in the pasture plots (III & IV) this age-class was not present at all. Number of juvenile individuals was by far highest in the forest (II) and not present at all in the pasture (III). Number of seedlings was ≥500 in all plots except for the pasture plot (IV), where it was exceptionally low. Adults of both sexes were generally better represented than juveniles and subadults, again except for the forest (II). The sex ratio was generally in favor of female plants, but only slightly so; here again the forest plot (II) stood out with more males than females. In forest plot (I) no males were present at all.

4.3.2 Use of leaves for thatch

The leaves of *P. aequatorialis* used for thatch (Fig. 4.2 A & B) are called *cade*. After harvest the leaves are split lengthwise and each half is folded 2–4 times into units 1 m wide called *tapas*. These thatching units are commonly marketed as bundles of 100, representing 4–5 stacks called *cestos* or *fardos*, each comprising 25 or 20 units respectively (Fig. 4.2 B). In Manabí, sales prices for leaves increased from 4–8 US$/*ciento* (100 units) in November 1992 to 30–50 US$/*ciento* in 2011, depending on quality, season and distance to the source. In 1990, a *P. aequatorialis* roofing for an average sized house (6 x 8 m, requiring 300–400 thatch units) cost 12–32 US$ which was much cheaper than corrugated iron roofing, which then cost more than 150 US$. In contrast, in 2011, the price for *P. aequatorialis* roofing for an average sized house cost from 90–200 US$, while the corresponding corrugated iron roofing cost about 200 US$. Corrugated iron lasts ca. 10 years, while a *P. aequatorialis* roofing is reported to last 5–6 years. The proportion of roofing based on natural materials decreased in the past 20 years: In 1990, along a 50 km stretch between El Carmen and Pedernales of 245 houses visible from the road, 49% had roofs with corrugated iron, 43% *Phytelephas*-thatch, 3% *Calathea*-thatch and 5% had mixed roofs. Of the houses with corrugated iron roofing many had kitchen niches with *Phytelephas*-thatch, because heat and smoke from fire places speed up the corrosion of corrugated iron. In 2012, a count along a 45 km stretch between Pueblo Nuevo and Pedernales showed that of 446 houses visible from the road, 97% had roofing made of corrugated iron or other synthetical material and only 3% had thatch of

Table 4.2 Population structure of *Phytelephas aequatorialis*

Population structure in five of the study plots in western Ecuador, giving number of individuals/hectare in each age-class and for adults divided into female/male. a and b are subdivisions of the agroforestry plot, b1 and b2 represent subdivisions of b. *Seedlings numerous but not counted. The population structure in plot V, acting as control for impact of leaf harvest on leaf and fruit production in Plot III was not recorded.

Plot	I Forest	II Forest	III Pasture	IV Pasture	VI Agroforestry				
Sub-plot					a & b	a	b	b1	b2
Area [m²]	1000	4608	875	5184	3642	1914	1728	918	810
Age class									
Adults (f/m)	40 (40/0)	365 (152/213)	365 (194/171)	131 (73/58)	373 (203/170)	387 (204/183)	359 (203/156)	207 (131/76)	531 (284/247)
Subadults	30	7	0	0	63	42	87	153	12
Juveniles	20	5007	0	142	55	94	12	11	12
Seedlings	1170	500	914	14	*	*	*	*	*

Table 4.3 Annual leaf production rates

(mean±SE) of 365 tagged *P. aequatorialis* individuals in six study plots (I–VI) in Ecuador (provinces Cotopaxi, Esmeraldas, and Manabí). In three plots (II–IV) both males and females were tagged, while in the forest (I), pasture (V), and agroforestry (VI) plots only female palms were included. Results from the agroforestry plot (VI) are presented according to low (≤3) and high exposure (>3) to sunlight. [a]Males produce more leaves than females (p≤0.0001). [b]correlation between light exposure and leaf production rate highly significant (Kendall: t=0.76, P≤0.0001). [c]lowland palms produce significantly more leaves than highland palms (p≤0.0001 for exposure >3, and p≤0.0001 for exposure ≤3).

Type of area	Forest	Forest			Pasture			Pasture			Pasture	Agroforestry		
Plot	I	II			III			IV			V	VI		
Region[c]	highland	lowland			highland			lowland			highland	lowland		
Sex/Exposure	Female	Male	Female	All	Male	Female	All	Male	Female	All	Female	Female ≤3	>3	All
#tagged palms	19	98	70	168	19	20	39	30	38	68	21	28	22	50
Leaves/year	4.1±0.1	6.9±0.2[a]	5.8±0.1[a]	6.4±0.1	6.9±0.2[a]	5.4±0.2[a]	6.1±0.2	9.1±0.4[a]	8.4±0.2[a]	8.7±0.2	6.1±0.2	8.2±0.4[b]	13.3±0.5[b]	10.4±0.5

Fig. 4.2 Raw materials and products

(A) *Phytelephas* roofing. (B) Stock-piled thatching units (*tapas*) manufactured with *P. aequatorialis* leaves. (C) Sequence of vegetable ivory discs (*animelas*) of different qualities: from brownish, yellowish (poor quality) to white in color (best quality). (D) Vegetable ivory seeds are sold to producers enclosed in the endocarp.

plants such as *Phytelephas aequatorialis* or *Carludovica palmata* Ruiz & Pav.

4.3.3 Leaf production

Leaf production rate varied from 4.1–13.3 leaves per year (Table 4.3) across study plots. The lowest leaf production rate was recorded in the highland forest plot (I) and the highest in the agroforestry plot in the lowlands (VI). Leaf production rates in the pasture plots (III–V) and the forest plots (I, II) were intermediate. In the agroforestry plot there was a highly significant positive correlation between light exposure and leaf production rate (Table 4.3). Leaf production rate also varied with gender and elevation. Males produce significantly more leaves than females and lowland palms produce significantly more leaves than highland palms (Table 4.3).

Harvesters were aware of the negative impact that leaf harvest may have on fruit production and that it can be mitigated by harvesting only fully developed leaves and leaving more leaves on female palms than on male ones. In Manabí in 1990 harvesters suggested that the four youngest leaves should be left on male palms and five on female palms, whereas farmers in Esmeraldas said that 5–7 leaves should be left on male palms, and that female palms should not be harvested at all. However, in 1990 both harvested male and female palms with only one or two leaves left were a common sight in both areas. In 2011, harvesters in Manabí suggested that 2–3 leaves should be left on the palms (one spear leaf and one opened leaf on male palms, one spear leaf and two opened leaves on female palms) when they are harvested, and this rule seemed to be followed according to our observations. In the pasture plot (III) an average of 4.6 ± 0.1 (3.0–7.0) leaves (n=39) were left after harvest. There was no significant difference (p=0.054) in number of leaves left on male and female palms (4.9 ± 0.2 and 4.3 ± 0.2, respectively). Harvesters told us that shade leaves are larger and of better quality than light-exposed leaves, which agrees with our observations. Average rachis length in the forest plot (I) was 743 ± 21 cm (n=18), and for the two pasture plots in the highlands (III, V) it was 536 ± 10 cm (n=60), i.e., a difference of more than two meters ($p\leq0.0001$). No significant difference in rachis length was found between male and female palms (p=0.56). Pinnae length neither varied between sexes nor between pasture and forest palms. Regarding leaf-quality, our observations confirm the harvesters' statements that many of the older leaves from exposed palms are in a poor condition, with a large percentage of dead pinnae, while shade palms generally have less worn leaves. In the agroforestry plot (VI) a lower total number of leaves in fully developed crowns was found and here the total number of leaves in the crown is significantly higher ($p\leq0.0001$) in palms growing under high light exposure compared to palms growing under low light exposure (26.5 ± 0.9 leaves for exposure >3 compared to 21.0 ± 0.8 leaves for exposure ≤3). This difference was also found in forest and pasture plot (I, III) in the highlands (14.7 ± 0.4 and 16.7 ± 0.5, respectively), though the difference was not significant (p=0.054). In the pasture plot (III), palms shed an average of 1.34 ± 0.22 (0.0–5.0) old leaves (n=39) during two years (1991–1993), indicating that full crown size is regained in approximately two years, if 4.6 leaves are left after harvest (Table 4.4).

Table 4.4 Mean functional life time for leaves

(±SE) of *P. aequatorialis* in western Ecuador and mean time (±SE) needed for regeneration of crowns if 4.6 leaves are left following leaf harvest in the forest, pasture, and agroforestry plots (I, III, VI). Results from the agroforestry plot (VI) are presented according to low (≤3) and high exposure (>3) to sunlight.

	I Forest	III Pasture	VI Agroforestry		
			Exposure		
			≤3	>3	All
Functional lifetime (years)	3.6 ± 0.1	2.8 ± 0.1	2.7 ± 0.1	2.1 ± 0.1	2.4 ± 0.1
Crown regeneration (years)	2.4 ± 0.1	2.0 ± 0.1	2.1 ± 0.1	1.7 ± 0.1	1.9 ± 0.1

4.3.4 Fruit and vegetable ivory production

The total duration from formation of the inflorescence bud to the mature infructescence was about 2.7 years in the lowlands (Table 4.5). The time for infructescences to reach full maturity varied from 2.1–7.5 years (Table 4.6). In the agroforestry plot (IV) in the lowland a highly significant correlation was found between exposure to light and production, whereas such correlations were absent in all other plots.

Infructescence production (i.e., infructescences shed per individual per year) in the lowlands was significantly higher than in the highlands. However, counting only individuals subjected to light exposure ≤3, production in the agroforestry plot (VI) did not differ significantly from production in the highland forest plot (I) with the same average exposure conditions. The number of new infructescences did not differ significantly among plots of the same region, but was significantly higher in the lowland than in the highland (Table 4.6). In total shade many individuals did not flower at all. Absence of flowering in shaded, caulescent individuals was pronounced in the agroforestry subplot b (Table 4.2).

In no cases had infructescences not present at time of tagging reached maturity before the end of the observation period (i.e., no inflorescences had passed through the whole cycle from flowering to mature infructescence within 723 days). The time needed for an infructescence to mature was generally longer in the highlands than in the lowlands (Table 4.6), however, the time needed from fertilization of ovules until the infructescence is large enough to be tagged (around 4–6 months) must be added to these values. Interestingly, infructescence

Table 4.5 Duration of flowering and fruiting stages

Duration of flowering and fruiting stages from the bud stage to the mature infructescence (mean±SE) of *Phytelephas aequatorialis* in two lowland plots (II, IV) in western Ecuador. Endosperm condition was determined by slashing a fruit.

	II Forest	IV Pasture
Flowering/Fruiting stage (females only)	Duration (days)	
1. Inflorescence bud	44.1±0.7	41.3±0.9
2. Inflorescence bud with emerged bracts	14.4±0.2	14.1±0.2
3. Fresh inflorescence	14.6±0.2	13.5±0.2
4. Withering inflorescence	14.8±0.1	14.5±0.2
5. Dry inflorescence	175.0±5.5	165.2±5.8
6. Infructescence(liquid endosperm)	168.0±1.3	163.8±1.5
7. Infructescence (incipient gelatinous endosperm)	32.2±1.1	30.1±1.0
8. Infructescence(hardened gelatinous endosperm)	56.0±0.6	58.8±0.9
9. Infructescence(soft opaque endosperm)	225.4±1.4	225.4±1.4
10. Mature infructescence(hard white endosperm)	253.4±1.8	250.6±2.8
11. Total duration of development (Inflorescence bud to mature infructescence)	997.9 (2.73 years)	977.3 (2.68 years)

Table 4.6 Development time for infructescences

Development time for infructescences (mean±SE) of *Phytelephas aequatorialis* individuals in six study plots (I–VI) in western Ecuador. Results from the agroforestry plot (VI) are also presented separately according to low (≤3) and high light exposure (>3) to sunlight. [a]corresponds to initially tagged number of infructescences per individual. [b]refers to the number of completely developed and shed infructescences per individual per year; significant difference between lowland and highland plots (p=0.005); highland plots: plot III significantly different to plot I and V (I: p=0.035 and V: p=0.0007, respectively). [c]calculated from the data for 459 days. [d]calculated from the data for 723 days. [e]refers to the number of infructescences produced per individual per year during the observation period; [f]highly significant difference between highland and lowland plots (p≤0.0001), no significant difference between plots of the same region. [f]refers to the number of infructescences present per individual at the end of the observation period, i.e., after 723 (I, III, V), 459 (VI), and 365 (II, IV) days, respectively. [g]represents the percentage of shed infructescences at the end of the corresponding observation period, i.e., after 723 (I, III, V), 459 (VI), and 365 (II, IV) days, respectively, that were present at time of tagging. [h]calculated by dividing number of tagged infructescences per individual with number of shed infructescences per individual per year. Figures in parenthesis correspond to total number of tagged infructescences. [i]highly significant correlation between exposure to light and production (Kendall: t=0.49, p≤0.0001). [j]no significant difference (p=0.14). [k]no significant difference in any of the plots (I–VI: p=0.83, 0.08, 0.35, 0.82, 0.1, and 0.89). [l]significant differences between number of new and shed infructescences were only found in the secondary forest plot (II) in the lowlands and the pasture plot (III) in the highlands (I–VI: p=0.52, 0.003, 0.65, 0.07, and 0.70). [m]significant differences between number of new and shed infructescences per individual per year.

Plot	Region	N	Tagged[a,d]	Shed/year[b,m]	New/year[e,m]	End[f,d]	Shed [%][g]	Development time (years)[h]
I Forest	highland	19	6.9±0.8 (97)	1.6±0.3[c,k] (21.7)	1.4±0.1[c] (20.2)	6.7±0.6 (94)	44	4.3
II Forest	lowland	70	5.8±0.6 (403)	2.8±0.3 (197)	5.1±0.1[c] (355)	8.0±0.4 (561)	49	2.1
III Pasture	highland	20	6.0±1.1 (120)	0.8±0.2[c] (16.1)	1.6±0.3[c] (32.3)	7.6±0.7 (152)	27	7.5
IV Pasture	lowland	38	13.3±0.6 (506)	4.4±0.3 (168)	4.6±0.2 (173)	13.4±0.6 (511)	33	3.0
V Pasture	highland	21	13.3±1.0 (279)	2.2±0.3[c] (47)	1.5±0.4[c] (31.8)	11.9±1.3 (249)	33	6.0
VI Agroforestry (all)	lowland	50	15.8±1.4 (791)	4.9±0.5[d] (246.5)	5.5±0.7[d] (275.9)	16.6±1.6 (828)	39	3.2
VI Agroforestry (≤3)	lowland	28	10.0±1.1 (281)	3.5±0.6[d,l,k] (98.6)	3.0±0.6[d] (83.5)	9.4±1.1 (262)	44	2.9
VI Agroforestry (>3)	lowland	22	23.2±1.8 (510)	6.7±0.8[d,l] (147.9)	8.8±0.9[d] (192.4)	25.7±1.9 (566)	37	3.5

development time was faster in the forest than in the pasture in highlands and lowlands, respectively, and faster in the shaded part of the agroforestry plot than in the more exposed part (Table 4.6).

Numbers of fruits per infructescence and seeds per fruit from two lowland plots (II, IV) did not differ significantly (Table 4.7). The overall seed weight per infructescence was similar in the forest and the pasture, but mean seed weight is significantly higher in the forest (Table 4.8). With a 1:1 sex ratio and 500 individuals per hectare the stand produced 750–2,500 kg vegetable ivory nuts (dry weight) per hectare per year in the highlands. Individuals in the highland plots produced more fruits per infructescence (25.1 vs. 21.2 and 18.9, respectively), whereas the number of seeds per fruit was higher in the lowland (6.0 and 5.8 vs. 4.9, Table 4.7 & 4.8). However, the overall mean number of seeds per infructescence ranged from 110–130 and did not differ significantly between the highlands and lowlands.

The harvest procedure was simple. When infructescences were mature they disintegrated and the fruits fell to the ground (Fig. 4.1 F) where the mesocarp was eaten by animals and the seeds, which are enclosed in the endocarp, could be collected. In forests with thinner populations collecting the seeds took more time and more seeds had been removed by animals according to the harvesters.

In times of high demand, it was common to harvest immature infructescences, which were buried or covered with leaves until decomposition made it easy to remove the remaining mesocarp. The resulting immature vegetable ivory seeds are called *tagua maceada* and are of poor quality because they typically are cracked, non-uniform in texture, and either transparent, or yellow to brown in color (Fig. 4.2 C). However, it was difficult to distinguish immature from mature vegetable ivory because the seeds were commonly sold enclosed in the endocarp (Fig. 4.2 D).

In the dense stands of *Phytelephas aequatorialis* in Esmeraldas and Manabí it was common practice to weed and clean the ground in a circle of 3–4 m around the stem, making it easier to locate and collect seeds. Palms were also occasionally cleaned from climbers (species of

Table 4.7 *Phytelephas* fruit from lowland plots

Data on number of fruits per infructescence of *Phytelephas aequatorialis* in western Ecuador and number of seeds per fruit from two lowland plots (II & IV). Data are mean values ±SE with sampling size in parenthesis. No significant differences detected. All weights are dry weight.

Plot	II Forest	IV Pasture
Fruits/infructescence	18.9±0.3 (42)	21.2±0.4 (37)
Seeds/fruit	5.8±0.1 (145)	6.0±0.1 (137)

Table 4.8 *Phytelephas* fruit from highland plots

Data on number of fruits per infructescence, number of seeds per fruit, weight and specific gravity of seeds, and endocarp weight of *Phytelephas aequatorialis* from the three highland plots (I, III, V) on the west Andean slopes in Ecuador. Data are mean values ±SE and with sampling size in parenthesis. All weights are dry weight. aFruits/infructescence multiplied with seeds/fruit multiplied with seed weight. bAverage number of infructescences per palm per year (Table 4.6) multiplied with seed weight per infructescence. cEach of these samples consist of 5–20 seeds. dmean seed weight is significantly higher (p≤0.0001).

	I Forest	III Pasture	V Pasture	All highland plots (I, III, IV)
Fruits/infructescence	25.2±1.4 (18)	25.6±2.7 (5)	24.9±1.2 (16)	25.1±0.9 (39)
Seeds/fruit	4.4±0.2 (97)	4.7±0.2 (65)	5.2±0.1 (285)	4.9±0.1 (447)
Seed weight (g)	39.4±0.9^d (266)	31.3±1.0 (76)	34.7±0.6 (268)	36.3±0.5 (610)
Seed weight/infructescence (kg)[a]	4.4	3.8	4.5	4.2
Seed production/palm/year (kg)[b]	7.0	3.0	9.9	6.6
Specific gravity of seeds (g/mm^3)[c]	1.34±0.02 (17)	1.36±0.02 (5)	1.35±0.01 (14)	1.35±0.01 (36)
Endocarp weight (g)	7.3±0.2 (228)	6.4±0.2 (76)	6.8±0.1 (267)	7.0±0.1 (571)

Clusia and *Ficus*) and litter from other trees was removed from the crowns. In the highlands (plot I) large litter loads partly covered the female inflorescences, but apparently with no impact as there was no significant difference in number of fruits per infructescence between palms from the forests and the pastures (I, III, and V, Table 4.8).

According to harvesters in Esmeraldas and Manabí thinning was done mainly by cutting non-productive male palms. The optimal sex-ratio, i.e., the minimum number of males per female without an adverse effect on fruit production, remains unknown, but harvesters suggested that a ratio of 1:3 (male/female) would be maintainable without resulting in a negative impact on fruit production.

Pest problems in the lowlands appeared to be limited even in the high density stands in Esmeraldas and Manabí. Partly or totally damaged seeds were only occasionally found here, but neither harvesters nor buyers considered it a problem. In the highlands, however, the bruchid beetle *Caryoborus chiriquensis* Sharp. lays its eggs into pores of the umbo and its´ larvae consume the endosperm. Infection rates can be very high, and the beetle represented an important obstacle to vegetable ivory exploitation in the highlands.

4.3.5 Impact of leaf harvest on fruit production

In the highland the number of infructescences cast during the observation period was significantly smaller in the harvested pasture (III) than in the forest and in the not-harvested pasture (Table 4.6), and also the weight of seeds/infructescence appeared to be lower (Table 4.8). Furthermore, in the harvested pasture (III) 5% of the tagged infructescences were damaged or aborted before reaching maturity. Regarding the development of new infructescences there was no significant difference found neither between pasture (III) and forest (I) nor between the two pasture plots in the highlands (Table 4.6).

4.4 Discussion

4.4.1 Distribution and population structure

Today vegetable ivory, the *tagua* represents one of the economically most important wild plant raw materials in Ecuador (Brokamp *et al.*, 2011). Exhaustive collection of vegetable ivory nuts may reduce regeneration, but in all exploited stands visited in Esmeraldas and Manabí our observations suggested that regeneration is not affected by the extent of performed collection of seeds (from the ground). Collection of mature seeds may even be intensified considerably without seriously impacting regeneration. Since higher light exposure increases fruit production in the lowlands some cutting of trees shading the palms may favor production. Seedling recruitment is also increased when the ground receives moderate amounts of sunlight compared to completely shaded ground.

In earlier booms destructive harvest was common (Acosta-Solis, 1944, 1948). Current exploitation is not likely to result in further depletion of remaining stands, since western Ecuador is now much more densely inhabited and most areas have owners who protect their palms. Also most forest has been cleared already and typically few palms are found in the remaining closed vegetation. In open pastures the complete absence of juveniles (plot III) and sub-adults (plots III and IV) suggests that the natural regeneration of populations located in this type of land management is not very succesful, which accordingly makes a continued existence of these populations unsecure. The sometimes extremely low number of early developmental stages as found for juveniles or seedlings in the pasture (plot III and IV) most probably is caused by a combination of both excessive exposure leading to dehydration, as has been reported for *Ceroxylon echinulatum* Galeano seedlings (Anthelme *et al.*, 2011) and destruction of seedlings and juveniles by cattle. This suggests that in order to ensure regeneration in pastures seedlings and juveniles have to be protected from cattle and excessive exposure to sunlight which could be done with meshed cages.

4.4.2 Leaf production and harvest

Higher leaf production in male individuals is beneficial to the farmers since they mainly harvest males leaving females to produce fruits. Higher leaf production in the lowlands (plots II, IV, VI) is probably due to higher temperature and higher and more evenly distributed precipitation. Also, the nutrient status of the alluvial soil on the periodically flooded banks of rivers (e.g., Río Santiago) is likely to be much better than that of the eroded pastures on the Andean slopes. However, Velásquez Runk (1998) reported lower leaf production rates in agroforestry plots that are frequently flooded than in drier plots.

Impact of leaf harvest on leaf production is limited. Although statistically significant, the difference between palms in harvested and not harvested pastures in the highlands (plots III, V) averaged only 0.7 leaves/palm/year. A linear correlation (leaf production vs. time) in the highland plots suggests that the number of harvestable leaves per year does not depend on whether leaves are harvested again before its crown has regained all its leaves or whether the harvest is postponed until the crown has reshaped completely. The time needed to regenerate the crown following harvest is important when harvest frequency is decided. High harvest frequency means more work per leaf harvested because the palm has to be climbed more often than at a low harvest frequency. In exposed palms a high harvest frequency may, however, be worthwhile because younger leaves of higher quality may be obtained. Another result of a high harvest frequency is that the continuously pruned crowns produce less shade than crowns which are left to reshape completely. In order to harvest a maximum number of leaves with minimum effort, individual palms thus should be harvested approximately every two years, as was suggested for *Ceroxylon echinulatum* (Duarte & Montúfar, 2012), exposed palms slightly more often than shaded ones.

4.4.3 Fruit production and development time

A lower fruit productivity in plots I, III, and V suggests that environmental factors, such as low night temperatures, lower precipitation, and/or lack of nutrients may represent the limiting factors in the highlands. In the lowlands (plots II, IV, VI), however, light appears to be the main limiting factor. For management purposes (e.g., in agroforestry systems) palms therefore should be placed in the canopy layer in the lowlands, whereas they may be grown in lower strata in the highlands.

Formation of the compact infructescence with an almost woody surface and very hard seeds is accompanied by extremely slow fruit development in *P. aequatorialis*. For the lowland plots II and IV we calculated infructescence development time using two methods. 1) We monitored the duration of the fenological stages in different inflorescences, which suggested that infructescence development lasted 2.7 years in both plots; 2) we divided numbers of tagged inflorescences with number of annually shed infructescences, which suggested a development time of 2.6 and 3.5 years, respectively. The second method of calculating development time assumes that the number of infructescences remains constant with time, i.e., the number of new infructescences matches lost infructescences. Though the observation period is relatively short, this condition appears to be fulfilled for plots I, IV, V and VI but not for plots II and III.

Destructive harvest by felling vegetable ivory palms, was common in the past (Acosta-Solis, 1944, 1948), but appears not to happen on an extensive scale any more. A more pungent threat to the palms comes from land conversion, e.g., from establishment of African oil palm plantations or pastures. *Phytelephas aequatorialis* seems to exemplify a rather sustainably exploited extractive resource. Nevertheless, the impact of vegetable ivory extraction on natural regeneration and on genetic diversity of the species, as well as the impact on the diverse associated fauna, should be monitored continuously.

Because some collectors harvest immature infructescences in times of high demand it may be meaningful to foster enrichment planting and seedling protection in pastures and mixed agroforestry-systems, or even to establish *P. aequatorialis* plantations in order to satisfy the high (and probably increasing) demand for vegetable ivory nuts. Introduction of batch coding with information on source (exact location), date of seed collection, and collector could help to identify collectors that supply mature vegetable ivory nuts to assure a more standardized and higher quality of raw material, which would be beneficial for all stakeholders in this commercial activity.

Caryoborus chiriquensis is typically found at 650–1500 m a.s.l. (Nilsson & Johnson, 1993). To reduce this pest problem fruits must be collected soon after falling, the mesocarp should be removed manually and the cleaned seeds need to be stored in closed containers. Because the beetle is unable to lay eggs into the pores before the mesocarp has been removed, entire fruits may be dried and shelled hereafter (i.e., cracking both mesocarp and endocarp at once), avoiding the laborious work of removing the fresh mesocarp (Borgtoft Pedersen, 1995).

4.4.4 Impact of leaf harvest on fruit production

Leaf harvest considerably reduces vegetable ivory production, even when enough leaves are left on female palms. In small highland communities or in the inlands of Esmeraldas there is no market for leaves for thatch. In touristic coastal regions (e.g., in Canuto, Manabí), however, hotels, bars, and other tourist facilities today are predominantly constructed in traditional architecture. Due to high prices, a short value chain, and considerable demand for *Phytelephas* thatch, the commercialization of leaves, hence, represents a more lucrative business for farmers in these areas. Shift from commercialization of vegetable ivory nuts to leaves for thatch may result in unsatisfied demand and rising prices of raw material in the national *tagua* industry and may further lead in resource limitation for companies exporting vegetable ivory discs (*animelas*). Measures should be taken to ensure sustainable use and commercialization of the two partially exclusive and locally competing products. Employ-ment of unsustainable practices in the harvest of seeds and leaves, decline of populations in pastures, and the resulting resource limitation in export of vegetable ivory discs, represent the main issues to be addressed to foster sustainable use of this valuable palm species in the future.

5   Parasitism and haustorium anatomy of *Krameria lappacea* (Dombey) Burdet & B.B.Simpson (Krameriaceae), an endangered medicinal plant from the Andean deserts

5.1   Introduction

Krameria lappacea (Domb.) Burdet & B.B.Simpson is a shrub of the semi-desert in the Andes of Peru (Fig. 5.1 A–C), southern Ecuador, and northern Argentina, Chile and Bolivia, occurring at elevations from sea level to 3600 m a.s.l. (Simpson, 1989a; Simpson *et al.*, 2004). In its native habitat *K. lappacea* is by far the largest root parasite, with individual shrubs reaching a diameter of 1.5 m and a height of up to 2 m. Inmost semi-desert areas of Peru it is also clearly the most common root parasite. However, the plant is becoming increasingly rare and is already commercially extinct in part of its range: Its roots have been considered as medicinal since pre-Colombian times (Daems, 1981; Frohne, 2006; Simpson, 1991) and the neolignans and tannins of the root cortex are the likely source of its pharmaceutical effect (Bussmann *et al.*, 2010; Carini *et al.*, 2002; Silva *et al.*, 2001; Tiemann *et al.*, 2007). The roots are sold under the name *Rhatany* or *Ratanhia* and are traded both locally and nationally, and exported to a considerable extent. Main exporting country is Peru, with an annual export of ca. 30 metric tons, all of which is destructively harvested from wild stands (Fig. 5.1 I–L). This corresponds to the annual destruction of ca. 140,000–395,000 plants for the export alone (Weigend and Dostert, 2005). Very little is known about its biology in spite of the extensive and highly destructive wild harvest. This is particularly problematic since *Krameria* is an ecologically important species at least in part of its natural range: it provides cover for a range of vertebrates (Szaro and Belfit, 1986, 1987) and is an important and highly palatably forage plants for a wide range of wild and domestic mammals (Anthony,1976; Anthony and Smith,1977; Bernard and Brown,1977; Hanley and Brady, 1977; Hayden, 1966; Miller and Gaud, 1989; Rautenstrauch *et al.*,1988; Szaro and Belfit,1986, 1987; Urness, 1973; Urness and McCulloch, 1973; Webb and Stielstra, 1979). As a consequence, *Krameria* is often heavily browsed and populations are very slow to recover from heavy browsing (Brown, 1950; Goldberg and Turner, 1986). Even more importantly, they are food plants for highly specialized oil-collecting bees (Gimenes and da Lobão, 2006; Simpson, 1989b; Simpson *et al.*, 1977).

K. lappacea belongs to the Krameriaceae Dumort, a monogeneric family of 18 species native to the deserts and semi-deserts of the Americas (Simpson, 1989a). The family belongs to the eurosid I clade (Soltis et al., 2000) and is now recognized as sister to Zygophyllaceae

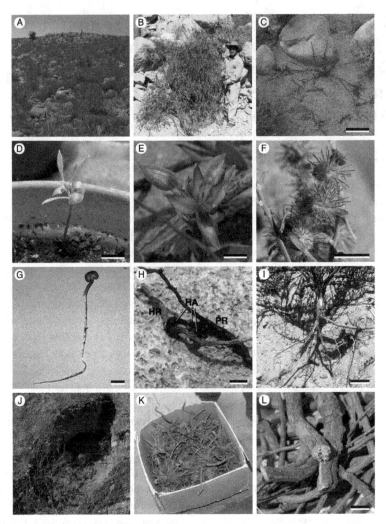

Fig. 5.1 Habitat and habit of *Krameria lappaceae*

(A) Habitat near San Antonio, Arequipa (Peru). (B) Mature individual near Chuquibamba, Arequipa, with one of the commercial harvesters, José Dionicio Inga Huamaní [Peru, voucher: M. Weigend *et al.* 9277 (B, HUSA, M, USM)]. (C) Juvenile plant near San Antonio, Arequipa (Peru). Scale bar = 10 cm. (D) Seedling in the green house at FU Berlin. Scale bar = 1 cm. (E) Open flower and buds. Scale bar = 1 cm. (F) Globose fruits with red, barbed spines. Scale bar = 1 cm. (G) Root of seedling from cultivation. Scale bar = 1 cm. (H) Haustorial connections with a root of *Balbisia verticillata*. Scale bar = 1 cm. (I) Excavated root system of mature plant at San Antonio. Scale bar = 10. (J) Typical excavation hole from harvested roots near Balsas (Cajamarca, Peru). (K) Unprocessed *Ratanhia* roots from a marketed in Tarma (Junín, Peru). (L) Dried and chopped *Ratanhia* roots on a market in Arequipa (Peru). Scale bar = 1 cm. Abbreviations: HA, haustorium; HR, host root; PR, parasite root.

in the order Zygophyllales (Angiosperm Phylogeny Group, 2009). *K. lappacea* has the typical zygomorphic oil-flowers (Fig. 5.1 E; Simpson, 1982) and one-seeded, glochidiate fruits (Fig. 5.1 F) of the genus. The biology of *K. lappacea* is of particular interest, since there is good reason to assume that it might be parasitic: *Krameria bicolor* S.Watson was the first taxon for which parasitism was documented (under the synonym *Krameria canescens* A. Gray; Cannon, 1910). Since then a total of seven species of Krameria have been studied and all were found to be hemiparasites (= semiparasites; Musselman, 1975; Simpson, 1989a, 2007). Hemiparasitic plants (in contrast to holoparasitic plants) are characterized by the possession of chlorophyll, still performing photosynthesis (Bresinsky *et al.*, 2008). Simpson (1989a, 2007) and Weber (1993) therefore suggested that the whole genus might consist of terrestrial hemiparasites, but data on the remaining 11 species have not been published. Simpson (1989a) reported that *Krameria* forms haustorial connections to a range of both herbaceous perennials and woody plants. The host plants reported in Simpson (1989a) for the three North American species of *Krameria* studied (*K. bicolor* S. Watson, *Krameria lanceolata* Torrey, and *Krameria erecta* Willd. ex Schultes) include a total of 15 angiosperm and two gymnosperm families, corresponding to a total of 21 species and genera, with 12 host species documented for *K. bicolor* alone. Even the formation of haustoria on their own roots (auto-parasitism) and on roots of neighbouring, conspecific individuals (intraspecific parasitism) has been reported.

Morphological and anatomical data have been published only for two North American species of *Krameria* (*K. bicolor* and *K. lanceolata*): The root system of these species consists of a short tap root and a series of sparsely branched lateral roots (Cannon, 1910; Kuijt, 1969; Musselman, 1975). The relative shortness of the tap root appears to be due to apical degradation of an originally longer primary root (Cannon,1910; Simpson,1989a). The individual lateral roots are covered with a thick and soft bark and are very flexible (Simpson, 2007). They radiate close to the soil surface (Cannon, 1910; Kuijt, 1969; Musselman,1975). The outer root cortex is dark red with a periderm that contains abundant tanniniferous material (Simpson, 1989a). The periderm originates in the outer layers of the secondary phloem and in older roots the outer cells of the periderm become suberized, filled with tannins and other substances (Musselman and Dickison, 1975). Cannon (1910) and Simpson (1989a) remarked on the apparent lack of root hairs on seedlings of *Krameria*.

Parasitism requires specialized organs called haustoria to tap the host plant (Kuijt, 1969) and *Krameria* is known to develop secondary haustoria, i.e., haustoria formed exclusively by lateral roots (Kuijt, 1969; Musselman, 1975). Two morphological regions of mature haustoria can be externally distinguished, the parent (mother) root and the swollen haustorium body. Anatomically, the parent root is similar to roots which do not bear haustoria and this is also true for the epidermis and cortex composing the outer part of the mature haustorium body, while the central parts of the haustorium are highly modified (Kuijt, 1969; Musselman,

1975; Musselman and Dickison, 1975). Progressing from the parent root through the hausto-rium towards the host root the following zones can be distinguished according to Musselman and Dickison (1975) and Simpson and Fineran (1970): (1) transition zone, (2) interrupted zone, (3) the vascular core, (4) the central parenchymatous core including the vessels (axial strands) that traverse it, (5) the portion of the haustorium situated within the host root (= endophyte). The transition zone is defined as the tissue containing the vascular connection between the xylem of the parent root and the vascular core of the haustorium body (Simp-son and Fineran, 1970). The region between the end of this transition zone and the vascular core is called interrupted zone (Simpson and Fineran, 1970), and consists of abundant pa-renchyma interspersed with scattered tracheary elements (Musselman and Dickison, 1975). The vascular core is a compact mass of xylem in the approximate centre of the haustorium, mainly consisting of vessel elements (ca. 95 %) plus some tracheids, fibres and parenchym-atic cells. The central parenchymatous core is defined by Musselman and Dickison (1975) as the portion of the mature haustorium bounded by the vascular core (proximally) and the host (distally). The central parenchymatous core is surrounded by a vascular cambium and the cortex in younger haustoria, respectively the periderm in older haustoria (Musselman and Dickison, 1975). The term endophyte (Kuijt, 1969) is used for the part of the haustorium that is situated inside the host root and consists only of parenchymatic cells and vessel ele-ments in mature haustoria (Musselman and Dickison, 1975). Initiated as a small intrusion of a few cells, the endophyte subsequently expands and becomes multi-lobed (Musselman, 1975). The vessel elements of the endophyte, which are continuous with the axial strands of the central parenchymatous core, are called terminal elements and provide the connection with the xylem cells of the host. The terminal elements differentiate from intrusive cells and develop thick, lignified walls and usually enter host xylem elements through lateral wall pits (Musselman and Dickison, 1975). Surprisingly, the overall anatomy of *Krameria* haustoria is also similar to secondary haustoria formed by other primitive terrestrial hemiparasitic species of distantly related orders such as Scrophulariales and Santalales, sharing, e.g., the characteristic haustorial tissues, with the development of a vascular core and the in general very similar xylem-system as well as the formation of an endophyte and the lack of phloem (Weber, 1993).

Only a few South American (Brazilian) species of *Krameria* are known to be parasitic (Simpson, 1989a) apart from the North American species, but no data on their host plants have been published, nor have parasitism or host plants been documented for western South American species. Morphological and anatomical data on the haustoria are available for only two North American species of this genus of 18 species. It is particularly remarkable that the only commercially important species, *K. lappacea*, has not been studied and that it is not even known whether it is at all hemiparasitic. However, an understanding of possible parasitism would be both crucial for an understanding of the ecological role that *K. lappacea* plays in its native habitat: Parasitic plants selectively reduce or promote the fitness of other plants and may have a massive indirect impact on both vegetation cover and quantitative and qualitative aspects of the vegetation itself (Callaway and Pennings, 1998; Joshi et al., 2000; Press and Phoenix, 2005; Pennings and Callaway, 1996; Watson, 2009). Selective removal of *Krameria* might thus affect other plants in its native habitat. Also, the range of host plants parasitized by *K. lappacea* this might also have important implications for its vulnerability (Marvier and Smith, 1997). Currently, *K. lappacea* is harvested to a great extent especially in Peru and by now it is commercially extinct in several parts of its range (Weigend and Dostert, 2005). Cultivation might be a way to provide raw material for the national and international markets, but this would require an understanding of its parasitic status and general biology.

The present study therefore aims at clarifying the parasitic status of endangered *K. lappacea* and, assuming parasitism, at identifying its host plants and characterizing its haustorial anatomy in comparison to the North American species studied elsewhere.

5.2 Material and methods

5.2.1 Study area and plant material

Field studies were conducted in 2003 in Peru, Department Arequipa, Prov. Arequipa. One study site (= Mollebaya) is situated just outside Mollebaya, towards the mountain "Lapacheta Chica", along a dry river bed, 2457 m a.s.l., S 16° 29.9090', N 071° 28.6280'. The other study site (San Antonio) is situated near San Antonio de Yarabamba, 3-4 km from Yarabamba, 2545 m a.s.l., S 16° 34.4110', N 071° 28.4560'. Both areas show relatively abundant populations of *Krameria* with a density of (0-)1-24 mature individuals of *Krameria* per 100 m2. Data on the climatic conditions of the study area in S Peru were recently published by Schwarzer et al. (2010). To confirm parasitism and to obtain haustoria for detailed study whole plants were carefully dug up and the secondary roots traced to the roots of their putative hosts. Most haustorial connections were undoubtedly lost, since the compacted and rocky soils exerted massive mechanical stress on the roots during excavation. *K. lappacea* and its host plants were vouchered and identified, vouchers were deposited in the Herbario de la Universidad San Augustin (HUSA) in Arequipa (Universidad Nacional de San Augustín, Arequipa, Peru). Host roots and the haustoria attached were cut out and preserved in AFA. Several seeds were germinated in the green house in the absence of host plants in order to study the early development of the root system.

5.2.2 Microscopic techniques

5.2.2.1 Preparation for light microscopy (LM)

The host root segments with the attached haustoria were trimmed and then cleaned for 3 min in an ultrasonic bath to remove soil and dust particles. The material was then dehydrated with a FDA fast dehydration series: 2 h FDA (fomaldehyde diethyl acetal, Merck AG; at room temperature = RT), 30 min FDA : isopropanol (2-propanol; at RT) 1:1, 2 h in isopropanol (at RT), then infiltrated for 2 h at 60 °C in tertiary butanol (3-butanol, 99.5 % extrapure, Acros Organics) : isopropanol 3:1, for 12 h at 60 °C or more in pure tertiary butanol, which was then gradually replaced in the course of ca. 48 h with Paraplast (Sherwood Medical; at 80 °C). Samples were then stored for 2-3 weeks in Paraplast to complete penetration. The mounted and trimmed objects were then sectioned into slices of 10-12 mm with a rotatory microtome (Leitz, 301-268) and transferred to microscope slides, that were left to dry on a heating plate (Medak Nagel GmbH; at 50 °C) for ca. 30 min. The Paraplast was then removed with xylol (Merck AG), the sections rehydrated through a descending, graded ethanol series, rinsed in water, and then stained using a combination of Safranine and Astra-Blue (Merck AG). Permanent mounts were then prepared by dehydrating the sections through an ascending graded ethanol series and sealing them in Eukitt (Kindler GmbH) using a cover slip. After 3 days of drying at 50 °C the slides were examined with a light microscope (Leitz Diaplan). Photographs were made with a digital camera for light microscope (DC 300 v2.0, Leica instruments GmbH).

5.2.2.2 Preparation for scanning electron microscopy (SEM)

Plant material was dehydrated with a FDA fast dehydration series (2 h FDA, 1 h FDA : acetone 1:1, 2 x 1h pure acetone), cleaned for 3 min in an ultrasonic bath and transferred again into fresh acetone. Complete drying of the plant material was carried out via critical point drying, following the protocol of Cohen and Shaykh (1973). After mounting and preparation of the plant material it was sputter-coated with gold as described by Echlin (1978) using a SCD 050 sputter coater (Oerlikon Balzers). Examination was performed using SEM (LEO 5430). The descriptive terminology used in this study follows Musselman and Dickison (1975).

5.3 Results

5.3.1 Hemiparasitism and host plants

The study shows that *K. lappacea* is indeed terrestrial hemiparasitic forming haustorial connections to several host plant taxa. The host plant taxa for which haustorial connections to *K. lappacea* were found are summarized in Table 5.1. A total of 106 haustoria were excavated at the two sites, most of them (86) at San Antonio. In spite of the technical difficulties of root excavation due to the hard and rocky ground, haustorial connections of the root systems of *Krameria* were found to host plants of 18 different species from 17 genera in 12 different plant families, mostly dicotyledonous plants. However, the only gymnosperm in the study

Table 5.1 List of host plants recorded in this study

direct haustorial connections found and corresponding voucher [all vouchers preserved at HUSA (Herbario de la Universidad San Augustin), collectors: Carazas = Richard Aguilar Carazas & F. Cáceres Humaní, Dostert = Nicolas Dostert & F. Cáceres Humaní

Host plant species	family	growth from	number of haustorial connections			host plant vouchers
			San Antonio	Mollebaya	total	
Ambrosia artemisioides Meyen & Walp. ex Meyen	Asteraceae	shrub	15	6	21	Carazas 420, Dostert 1005
Gochnatia arequipenses Sandwith	Asteraceae	shrub	5	-	5	Carazas 416, Dostert 1004
Grindelia tarapacana Phil.	Asteraceae	shrub	4	2	6	Carazas 459, Dostert 1001
Helogyne sp.	Asteraceae	shrub	3	-	3	Carazas 458
Balbisia verticillata Cav.	Ledocarpaceae	shrub	3	2	5	Carazas 422, Dostert 1022
Corryocactus aureus (Meyen) Hutchison	Cactaceae	shrub	2	-	2	Carazas 424
Cumulopuntia (Cumulopuntia) sphaerica Foerster	Cactaceae	shrub	6	3	9	Carazas 425
Opuntia (Cumulopuntia) corotilla Schumann ex Vaupel	Cactaceae	shrub	9		9	Carazas 426
Encelia canescens Lam.	Asteraceae	(sub-)shrub	7	1	8	Carazas 428, Dostert 1002
Ephedra americana Humb. & Bonpl. ex Willd.	Ephedraceae	shrub	4	-	4	Carazas 430, Dostert 1016
Huthia coerulea Brand	Polemoniaceae	shrub	1	-	1	Carazas 436, Dostet 1010
Juniella aspera (Gillies & Hook.) Moldenke	Verbenaceae	subshrub	3	-	3	Carazas 403
Mentzelia scabra Kunth *subsp. chilensis* (Gay) Weigend	Loasaceae	subshrub	2	-	2	Weigend 7466
Paronychia microphylla Phil.	Caryophyllaceae	perennial herb	8	1	9	Carazas 408, Dostert 1008
Portulaca pilosa L.	Portulacaceae	perennial herb	5	2	7	Carazas 415, Dostert 1019
Tarasa operculata (Cav.) Krapovickas	Malvaceae	shrub	3	1	4	Carazas 421, Dostert 1007
Eragrostis peruviana (Jacq.) Trin.	Poaceae	annual/subperennial herb	-	2	2	Dostert 1017
Verbena clavata Ruiz & Pav.	Verbenaceae	perennial herb/subshrub	6	-	6	Carazas 401
total: 18	12		86	20	106	

area, *Ephedra americana*, was also found to be parasitized by *K. lappacea*. The single most commonly documented host plant is *Ambrosia artemisioides*, which is also the dominant plant species at both study sites (ca. 1/5th of the haustorial connections found). Parasite roots formed one to several connections to the host root, in individual cases up to three haustoria in immediate vicinity to each other (Fig. 5.1 H). Several cases of auto-parasitism were observed.

5.3.2 Roots and root system

The young plants raised in the green house developed a primary root (Fig. 5.1 G), that soon branched to form perpendicular secondary roots, the roots were free of visible root hairs. The mature root system of *K. lappacea* consists of a rather short primary root and a couple (ca. 8–15, Fig. 5.1 I & K) of very flexible secondary roots radiating from it in a depth of 20–70 cm. They extend to a horizontal length for 1–3(–4) m (and possibly more). The primary root initially grows to a considerable depth (ca. 15–70 cm), but its distal part later (after the formation of haustoria) decays beyond the attachment of the secondary roots, and then has a length of typically 10–25 cm. The primary root reaches a diameter of ca. 25–50 mm in mature plants (Fig. 5.1 I). The mature secondary roots of K. lappacea are weakly branched, dark red in colour and have a smooth and soft bark (Figure 5.1 I), which becomes flaky once dry (Fig. 5.1 L). The laterals arising from the secondary roots were found to either terminate in haustoria or in slightly thickened, spindle-shaped root "tubers", here interpreted as pre-haustoria. The lateral roots examined in this study had a diameter of 0.3–2.0 mm. Only the secondary and lateral roots form haustoria (Fig. 5.1 H) and often terminate in them, due to degeneration of the parasite root beyond the haustorium.

The mature roots have a tanniniferous periderm consisting of 6-10 cell layers of collapsed phellem cells (cork), which are compressed in transverse section (Fig. 5.2 A–C). These are followed by the cork cambium and 3–5 cell layers of cortex parenchyma with cells also tanniniferous and tannins aggregated into irregularly rounded structures up to 20 mm in diameter (Fig. 5.2 C). Crystal druses were also present in the cortex parenchyma cells. Mature roots have a partly interrupted sclerenchymatous ring of 1–3 cell layers, secondary phloem is present in the form of scattered cell groups. An indistinct cambium is also visible. The xylem tissue in the centre of the root makes up approximately half the root diameter. The tracheae and vessels are wide, with diameters of up to 50 mm. Between the vessels are wood parenchyma cells and irregular medullary rays (Fig. 5.2 A–C).

5.3.3 Haustoria

The haustoria of *K. lappacea* usually are rounded, blackish structures up to 4 mm wide and 10 mm long. *Krameria* haustoria are formed on both primary and lateral roots of its hosts and are by definition secondary haustoria (Kuijt, 1969). Haustoria are mostly laterally attached to the host root, although some smaller host roots may be almost completely surrounded by them. The parent root of the haustoria is structurally similar to the lateral roots and terminates in the haustorium body. The haustorium body is a depressedly-globose structure much wider than the parent root. It forms a broad collar on the host root surface effectively obscuring the point of entry of the endophyte into the host root (Fig. 5.2 D). The mature haustorium body is histologically complex (Fig. 5.3 A). Periderm and cortex anatomically resemble that of the parent root, with abundant crystal druses in the cortex parenchyma (Fig. 5.3 B). The central portion of the haustorium body contains tightly bundled vascular tissue and represents the vascular core (Fig. 5.3 C), proximally (towards the parent root) bordering on the interrupted zone (Fig. 5.2 E). The interrupted zone consists of parenchyma cells scattered with vessels forming a connection between the vascular core and the parent root xylem. Distally, towards the host root, the central parenchymatous core extends to the contact zone with the host. This central parenchymatous core consists of smaller and thin-walled, tightly packed, isodiametric to elongated cells which are provided with conspicuously large nuclei (Figs. 5.2 F & I, 5.3 A). The tracheary elements of the central parenchymatous core represent the axial strands forming the xylem connection between the vascular cylinder of the host and the vascular core of the haustorium (Fig. 5.3 E). The distal part of the central parenchymatous core, which is embedded in the host root tissue (Fig. 5.3 F), is called the endophyte (Figs. 5.2 G & H, 5.3 G). The endophyte here is conspicuously lobed (Fig. 5.2 F). The vessel cells of the axial strands that are situated in the endophyte tissue are the terminal elements of the parasite xylem and provide the connection to the host xylem (Fig. 5.2 I). Phloem cells were not recognizable in any of our sections in any part of the haustoria.

5.3.4 Pre-haustoria

Some lateral roots were found to terminate in spindle-shaped structures 1–5 mm wide and 2–6 mm long (Fig. 5.2 J, K). These structures are here interpreted as pre-haustoria. They are thicker than the adjacent part of the parent root, and the apex has an often fragmented bark cortex (Fig. 5.2 L), which is likely where the developing haustorium penetrates the host root. These structures are anatomically similar to the lateral roots, but show a more voluminous parenchyma with up to 9 layers of thin-walled, more or less rounded cells, which is responsible for the wider diameter of the root tubers. This parenchyma appears to fall into two distinct regions, a central part which is continuous with the parenchyma of the parent root, and a distal, outer cap (Fig. 5.2 L). Crystal druses were observed in the inner parenchymatic cells, but not in the outer cap. The periderm of the root tubers is less extensive than in the lateral roots, and organized as one layer of cork cambium and 4–5 cell layers of (cork) phellem cells (Fig. 5.2 J–L). Tannins appear to be less abundantly present than in the lateral roots. The cambium, surrounding the xylematic central core, consists of up to 3 cell layers. The vascular cylinder has a diameter of 0.1–0.2 mm and is less well developed than in the lateral roots.

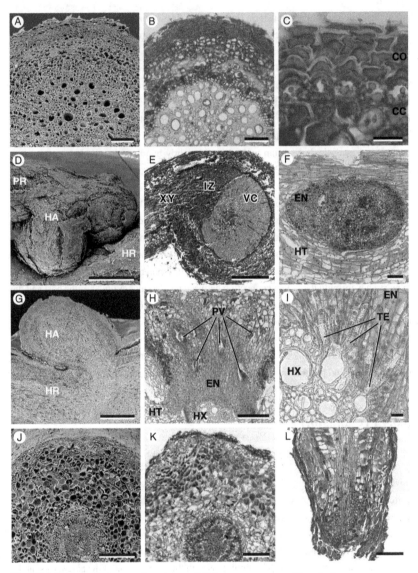

Fig. 5.2 Micromorphology and anatomy of *K. lappacea* roots and haustoria

(A) Transverse section through secondary root (SEM). Scale bar – 100 mm (B) transverse section through secondary root (LM). Scale bar – 100 mm (C) Detailed view of root periderm (LM). Scale bar – 25 mm. (D) Haustorium on a root of *Grindelia tarapacana* (SEM). Scale bar – 1 mm. (E) transverse section through the haustorium and longitudinal section through its parent root (LM). Scale bar – 500 mm (F) transverse section through endophyte (and longitudinal section through the parasitised root of *Ambrosia artemisioides*) (LM). Scale bar – 100 mm. (G) Section through an auto-parasitizing haustorium (SEM). Scale bar – 1 mm. (H) Longitudinal section through haustorium on root of *A. artemisioides* (LM). Scale bar – 200 mm (I) Detailed view of central parenchymatous core with axial strands connected to the root xylem of host (*A. artemisioides*) (LM). Scale bar – 20 mm (J) transverse section through a pre-haustorium of *K. lappacea* (SEM). Scale bar – 500 mm (K) transverse section through a pre-haustorium (LM). Scale bar – 500 mm. (L) Longitudinal section through (the tip of) a pre-haustorium (LM). Scale bar – 200 mm. Abbreviations: CC, cork cambium; CO, cork; EN, endophyte; HA, haustorium; HR, host root; HT, host tissue; HX, host xylem; IZ; interrupted zone; PR, parent root; PV, parasite vessel; TE, terminal element (of axial strand); VC, vascular core; XY, xylem (of the parent root).

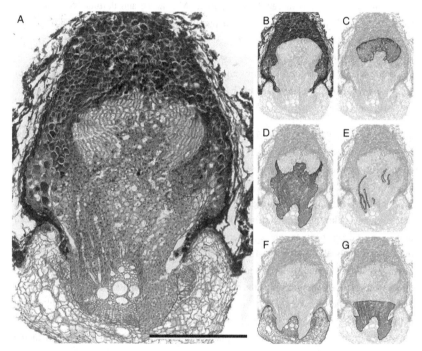

Fig. 5.3 Haustorium anatomy of *K. lappacea*

longitudinal section, LM. (A) whole haustorium view. Scale bar – 500 mm (B–G) highlighted are the different regions/elements of the mature haustorium, (B) cortex and periderm (the mantle), (C) vascular core, (D) central parenchymatous core, (E) axial strands (and terminal elements) connecting host and parasite xylem, (F) host root tissue (transverse section), (G) endophyte (the part of the central parenchymatous core that penetrates the host tissue).

5.4 Discussion

It has been claimed that all species of *Krameria* are hemiparasitic based on the observation of hemiparasitism in 7 of the 18 species recognized (Simpson, 1989a, 2007; Weber, 1993). This assumption can be clearly confirmed for *K. lappacea*: we documented a large number of haustoria on the roots of *K. lappacea* and the absence of both fine roots and root hairs in young plants. These findings indicate obligate hemiparasitism for the species. Compared to published data on *K. erecta, K. bicolor* and *K. lanceolata* (Simpson, 1989a) we could identify an even wider range of host plants for *K. lappacea*, including an additional four angiosperm families (Caryophyllaceae, Ledocarpaceae, Malvaceae and Portulacaceae) as hosts for *Krameria* and confirming other families previously documented. Another 18 host plant species are here documented for *Krameria*, nearly doubling the number from the previously known 23 to 41. *Krameria* parasitizes a very broad range of host plants, including herbaceous and woody taxa and angiosperms and gymnosperms. Further studies in other parts of the natural range of *K. lappacea* will likely reveal additional host species, since few if any of the host plants here recorded are present throughout the range of *K. lappacea*.

The hemiparasitic nature of *Krameria*, that is documented here for the first time, has to be seen in light of the fact that the plants can reach a considerable size and are naturally (i.e., on those areas where they have not been commercially collected in the past) very common. Recent studies clearly indicate that parasitic plants may play a crucial role in maintaining biodiversity (Callaway and Pennings, 1998; Joshi *et al.*, 2000; Press and Phoenix, 2005; Pennings and Callaway, 1996; Watson, 2009). This may be particularly true for *K. lappacea* in its (semi-) desert habitats, where a relatively small number of other perennial species are present and likely all of them are affected by *K. lappacea* in their ecological performance, possible leading to reduced vigour and cover, as shown for other root parasites (Joshi *et al.*, 2000). A considerable proportion of the plant diversity in these habitats is found in the annual and ephemeral species (Weberbauer,1945), which would be promoted in their development by a selective suppression of perennial and shrubby species. Uncontrolled wild harvest of the species may thus negatively affect the plant communities and their biodiversity, with the effects reaching well beyond the numerous animal species directly dependent on *Krameria*. There are copious data on animal species dependent on *Krameria* for food or cover, primarily from North America (Anthony, 1976; Anthony and Smith, 1977; Bernard and Brown, 1977; Brown, 1950; Gimenes and da Lobão, 2006; Goldberg and Turner, 1986; Hanley and Brady, 1977; Hayden, 1966; Miller and Gaud, 1989; Rautenstrauch *et al.*, 1988; Simpson, 1989b; Simpson *et al.*, 1977; Szaro and Belfit, 1986, 1987; Urness, 1973; Urness and McCulloch, 1973; Webb and Stielstra, 1979). Detailed studies from South America would be desirable for a better understanding of the ecological importance of this endangered species.

K. lappacea shows rather primitive characters for a terrestrial hemiparasite in terms of specialization, such as the exclusive formation of secondary haustoria, lack of organ reduction (stems and leaves normally developed), large seeds, low host-specificity and the ability to germinate in the absence of a host Weber (1993). Cultivation of *Krameria* to produce the *Rhatany*-roots for the national and international trade will be complicated by the need to cultivate it together with at least one of its host plants. On the other hand, it seems unlikely that the availability of host plants is a limiting factor for the natural distribution and abundance of *K. lappacea*, since it appears to be very unspecific in its host plant choice. Morphology and anatomy of *K. lappacea* haustoria closely resemble those described for North American congeners as described by Cannon (1910, 1911) and Musselman (1975). Similarity covers all main aspects of haustorium histology, e.g., presence of transition zone, interrupted zone vascular core, the central parenchymatous core including the vessels (axial strands) and lobed endophyte, tanniniferous cork, but also details such as the presence of crystal druses in the periderm and the parenchyma (Musselman, 1975). Haustorial structure in *Krameria* thus appears to be relatively conserved across the genus.

K. lappacea is an important source of income for part of the rural population, especially in Peru. Conservation and adequate management for the species should therefore have high priority. The data here presented provide a first step towards a biological understanding of this overexploited and increasingly rare species. Unfortunately, there are no published data on the abundance of *K. lappacea* across its range and nothing is known about its growth rates or natural recruitment. Future studies should be directed towards an improved understanding of these factors in order to formulate strategies for its conservation and sustainable use.

6 Now, where did all the *Rhatanies* go? Abundance, seed ecology, and regeneration of *Krameria lappacea* from the Peruvian Andes

6.1 Introduction

Krameria lappacea (Dombey) Burdet & B.B.Simpson is a slow-growing shrub from Andean semi-deserts of Peru, Southern Ecuador, Northern Argentina, Chile, and Bolivia. It is an obligate root hemiparasite parasitizing a wide range of flowering plants (Brokamp *et al.*, 2012; Simpson, 2007). The zygomorphic oil-flowers develop into large, one-seeded, glochidiate fruits typical of the genus (Simpson, 1982, 2007). *Krameria* seeds germinate happily in the absence of a host plant (Kuijt, 1969), but plantlets need the connection to a host within a short period of time in order to survive (Musselmann, 1996). Optimal temperature for germination is reported to be 24 °C (for *K. lanceolata*, Musselmann and Mann, 1977). For species of several plant families that dwell in arid environments it is reported that seed burial into the upper soil layer ($\leq$5 cm depth) significantly enhances seed germination (Beck and Vander Wall, 2010; Froud-Williams *et al.*, 1984; Martínez-Duro, *et al.*, 2009; Ren *et al.*, 2002; Vander Wall, 1993; Waitman *et al.*, 2012), however, in *Krameria lappacea* the effect of seed burial on germination success has not been investigated so far.

Krameria lappacea represents a well-known medicinal and dye plant with various reported traditional uses (Bussmann *et al.*, 2010, 2011; Bussmann and Sharon, 2006; De la Cruz *et al.*, 2005, 2007; Hilgert, 2001; Núñez *et al.*, 2009; Thomas *et al.*, 2009a). Root extracts of *K. lappacea* have been considered as medicinal since pre-Colombian times (Daems, 1981; Simpson, 1991). Pharmacological properties are generally centred around the astringent and anti-inflammatory properties of its tanniniferous root extracts (Simpson, 1989a, 1991). Over 200 years ago *K. lappacea* was introduced into European medicine as *"ratanhiae radix"*, resulting from the advocacy of its use by Hipólito Ruiz (Daems, 1981; Ruiz, 1797, 1799; Simpson, 1991), and over the years found its way into many pharmacopoeias (Komission E, 1989). Today *Krameria* root extracts are used in a range of cosmetic and pharmaceutical preparations, particularly in oral care products (Adwan *et al.*, 2012; Grazi, 2008) due to their antioxidant, antifungal, and antimicrobial activities (Artini *et al.*, 2012; Carini *et al.*, 2002) and have gained high-profile as promising ingredients in several cosmetical and pharmaceutical preparations (Adwan *et al.*, 2012; Artini *et al.*, 2012; Bussmann *et al.*, 2010, 2011; Carini *et al.*, 2002; Tiemann *et al.*, 2007). Among verifiable active compounds are tannins (catechins and proanthocyanidins), lignans, and neolignans (Artini *et al.*, 2012; Baumgartner *et al.*, 2011; Carini *et al.*, 2002; Scholz and Rimpler, 1989).

The roots are widely sold (under the name *Rhatany* or *Ratanhia*), locally to internationally, with no data available on the overall *Rhatany* harvest in Peru. *Rhatany* is destructively harvested from natural populations and commercial cultivation is unknown (Weigend and Dostert, 2005). Despite its traditional (domestic) use over its entire distributional range, commercial sourcing mainly takes place in Peru (Weigend and Dostert, 2005), probably because of a comparatively lower abundance in the other countries such as in, e.g., Bolivia (c.f., Thomas et al., 2009b). Between 2000 and 2004 Peru exported an average of 33 tonnes of *Rhatany* roots per year. Due to extensive commercialization and the impossibility of commercial cultivation the wild stands of this plant are becoming increasingly rare. Consequently, it is already threatened and commercially extinct in part of its range, particularly in areas on which wild collection was focussed in the past, i.e., Lima and the central Peruvian departments Junín, Ayacucho, and Apurímac (Brokamp *et al.*, 2012; De la Cruz. *et al.*, 2007; Weigend and Dostert, 2005). Accordingly, *K. lappacea* is already listed as critically endangered in close vicinity to highly populated areas (i.e., in Canta, Lima; De la Cruz et al., 2005, 2007).

In 2001, a German cosmetics and pharmaceutical company, that produces dental care products from *Krameria* root extracts (Weleda AG), initiated a project to investigate the possibilities for sustainable wild harvest of Rhatany in Peru in collaboration with the GTZ (Deutsche Gesellschaft für Technische Zusammenarbeit, now part of GIZ – Gesellschaft für Internationale Zusammenarbeit), INRENA (Instituto Nacional de Recursos Naturales, now part of Minag – Ministerio de Agricultura, Ministry of Agriculture of Peru), and botconsult GmbH. Only few data on the reproduction biology, current distribution and abundance of *Krameria* in Peru had been published. Therefore, extensive investigations were undertaken in order to establish a basis for a shift from over-exploitation towards a sustainable management of this resource (Weigend and Dostert, 2005). The data here presented include abundance data from several regions in Peru (Dept. Arequipa – San Antonio & Chuquibamba, Dept. Ancash – Caraz, Dept. Cajamarca & Amazonas – Balsas, Dept. Junín – Tarma, Dept. Moquegua – Omate) and the current state of *K. lappacea* populations regarding their natural rejuvenation. Furthermore, we provide experimental results on *Ratanhia* germination, which are relevant to the management options available.

6.2 Material and Methods

6.2.1 Study areas

Field studies were carried out between 2004 and 2012 at overall 19 sites in the Peruvian Andes at altitudes ranging from 880 to 3050 m above sea level (in 6 locations within the departments Amazonas, Cajamarca, Ancash, Junín, Moquegua, and Arequipa). Subsistence agriculture, livestock breeding, and extraction of plant resources from the wild are among the main activities of locals in these areas. The type of and intensity of *Ratanhia* harvest differs dramatically between the different study areas: In Caraz, harvest is only for the local market, with very low overall intensity. In San Antonio there is both harvest for the local and regional (Arequipa market) and for the purposes of export, with the collection area very well accessible by road. Balsas has been subject to massive commercial collection, starting between 1998 (when the populations were still virtually untouched) and ca. 2005. For a while, it supplied the bulk of the *Ratanhia* exported to, e.g., Germany. Similarly, *Ratanhia* was still very abundant in Omate 2006, but was then virtually destroyed by illegal harvest for the smuggling to Bolivia (from where it was sold as "cultivated in Bolivia" to Germany). In Tarma, there is a flourishing local trade and also transport to Lima, since the plant is there sold on the city markets. Chuquibamba is a very special case, since it is a fairly inaccessible area, and *Ratanhia*-collection is in the hands of essentially one family, which supplies to market in Arequipa and exports to some degree. Here, sustainable harvest techniques are traditionally employed in order to ensure the continuous regeneration of the populations. Detailed information on all study locations is provided in Table 6.1.

6.2.2 Species

Krameria lappacea (Domb.) Burdet & B.B.Simpson (*Ratanhia* or *Rhatany*) is a slow-growing shrub of the semi-deserts in the Andes of Peru, southern Ecuador, northern Argentina, Chile and Bolivia, occurring at elevations from sea level to 3600 m above sea level. It belongs to the Krameriaceae Dumort, a monogeneric family of 18 species native to the deserts and semi-deserts of the Americas (Simpson, 1989a). Ecologically *K. lappacea* is of particular interest because it is a highly generalistic, but obligate terrestrial hemiparasite providing forage for a wide range of wild and domestic animals (see Brokamp *et al.*, 2012); highly specialized bees of the genus *Centris* collect oil from the elaiophores within its zygomorphic flowers, which develop into striking one-seeded, glochidiate fruits (Simpson *et al.*, 1977; Simpson, 1982).

6.2.3 Methods

6.2.3.1 Precipitation rates in the study areas

Data on monthly precipitation rates of the past five years were obtained from the SENAMHI website (Servicio Nacional de Meteorologia e Hidrologia; SENAMHI, 2013) and the five year means of monthly and annual precipitation rates were calculated for the areas under study (except for Balsas, where some data was missing, resulting in a four year mean only). Data used came from the closest possible meteorological stations from roughly the same altitudes as the study locations are situated (Table 6.2).

Table 6.1 Detailed information on all study locations

Origin and number of acquired inventory data, extent of wild collection, data collectors and time of data collection.

Study area	Location	Extent of wild collection	No. of inventories (100 m²)	Field work
Balsas 1	Department Amazonas, Province Chachapoyas, Locality: along the road from Balsas to Chachapoyas, 2 km SE of Balsas, 06°51'22.4''S, 078°00'26.4''W, 880 m.	Intensive commercial plant collection in the past.	16	Henning & Brokamp, May 2006
Balsas 2	Department Cajamarca, Province Celendin, Locality: along the road from Celendin to Balsas, 6 km W of Balsas, 06°50'49.6''S, 078°02'20.9''W, 1176 m.	Intensive commercial plant collection in the past.	15	Henning & Brokamp, May 2006
Caraz 1	Department Ancash, Province Huaylas, Locality: 5 km after Caraz, north of the road to Casma, 08°59'54.1''S, 077°49'24.9''W, 2200 m.	Domestic plant collection only.	15	Weigend, Henning, Schwarzer & Brokamp, April 2006
Caraz 2	Department Ancash, Province Huaylas, Locality: 5 km after Caraz, south of the road to Casma, 08°59'54.1''S, 077°49'24.9''W, 2200 m.	Domestic plant collection only.	7	Weigend, Henning, Schwarzer & Brokamp, April 2006
Tarma 1	Department Junin, Province Tarma, Locality: north of the road from Tarma to La Merced, close to Vilcabamba, 11°21'24''S, 075°37'15.6''W, 2912 m.	Intensive commercial plant collection in the past.	17	Henning & Brokamp, May 2006
Tarma 2	Department Junin, Province Tarma, Locality: south of the road from Tarma to La Merced, close to Vilcabamba, 11°21'24''S, 075°37'15.6''W, 2912 m.	Intensive commercial plant collection in the past.	14	Henning & Brokamp, May 2006
Omate	Department Moquegua, Province General Sanchez Cerro, Locality: along the road from Omate to Tamina, 16°33'19.7''S, 070°59'44.8''W, 2540 m.	Intensive commercial (marauding) plant collection in the past.	10	Ackermann, December 2006

Table 6.1 Detailed information on all study locations (continued)

Origin and number of acquired inventory data, extent of wild collection, data collectors and time of data collection.

San Antonio A1	Department Arequipa, Province Arequipa, Locality: close to San Antonio de Yarabamba, on the western foot of the hills, just outside the extractive reserve established by INRENA and the WELEDA AG in 2003, 16°34'56''S, 071°27'19''W, 2648 m.	Intensive commercial plant collection in the past.	5	Weigend & Aguilar, April 2004
San Antonio A2	Department Arequipa, Province Arequipa, Locality: close to San Antonio de Yarabamba, eastern slope and crest of the hill, inside the extractive reserve (see San Antonio A1), 16°35'18''S, 071°27'31''W, 2699 m.	Intensive commercial plant collection.	23	Weigend & Aguilar, April 2004
San Antonio A3	Department Arequipa, Province Arequipa, Locality: close to San Antonio de Yarabamba, western slopes, inside extractive reserve (see San Antonio A1), 16°36'36''S, 071°26'53''W, 3012 m.	Intensive commercial plant collection.	12	Weigend & Aguilar, April 2004
San Antonio B1	Department Arequipa, Province Arequipa, Locality: close to San Antonio de Yarabamba, , inside the extractive reserve (see San Antonio A1), hillside of Mount Pedregoso, west of Quebrada Cachihuasi, 16°34'38.9''S 071°27'18.2''W, 2450-3050 m.	Intensive commercial plant collection.	18	Aguilar, May-November 2004
San Antonio B2	Department Arequipa, Province Arequipa, Locality: close to San Antonio de Yarabamba, , inside the extractive reserve (see San Antonio A1), hillside of Mount Santa Catalina, between Quebradas Abra Grande and El Huarangal, 16°34'53.3''S, 071°27'19.6''W,2450-3050 m.	Intensive commercial plant collection.	18	Aguilar, May-November 2004
San Antonio B3	Department Arequipa, Province Arequipa, Locality: close to San Antonio de Yarabamba, , inside the extractive reserve (see San Antonio A1), hillside of Mount Santa Catalina, between Quebradas El Huarangal and La Despachana, 16°34'33.2''S, 071°27'16.4''W, 2450-3050 m.	Intensive commercial plant collection.	18	Aguilar, May-November 2004

Table 6.1 Detailed information on all study locations (continued)

Origin and number of acquired inventory data, extent of wild collection, data collectors and time of data collection.

Chuquibamba 1	Department Arequipa, Province Condesuyos, Locality: Chuquibamba, Zona de Arequipilla, Quebrada Los Sunchos, 15°54'06''S, 072°36'27''W, 2527 m.	Privately controlled commercial plant collection.	10	Caceres, April-May 2012
Chuquibamba 2	Department Arequipa, Province Condesuyos, Locality: Chuquibamba, Quebrada de Pampas de Asia, 15°53'11''S, 072°37'55''W, 2650 m.	Privately controlled commercial plant collection, area harvested in 2011.	10	Caceres, April-May 2012
Chuquibamba 3	Department Arequipa, Province Condesuyos, Locality: Chuquibamba, Pacaychacra, 15°53'34''S, 072°34'22''W, 1722 m.	Privately controlled commercial plant collection.	10	Caceres, November 2011
Chuquibamba 4	Department Arequipa, Province Condesuyos, Locality: Chuquibamba, Arequipilla Baja, 15°51'28''S, 072°38'03''W, 2445 m.	Privately controlled commercial plant collection.	10	Caceres, November 2011
Chuquibamba 5	Department Arequipa, Province Condesuyos, Locality: Chuquibamba, Quebrada Honda, 16°11'47''S, 072°49'04''W, 1850 m.	Privately controlled commercial plant collection.	5	Caceres, October 2010
Chuquibamba 6	Department Arequipa, Province Condesuyos, Locality: Chuquibamba, Quebrada de Pachana, 16°04'37''S, 072°43'42''W, 1688 m.	Privately controlled commercial plant collection.	5	Caceres, October 2010
Total inventory number			238	

Table 6.2 Meteorological stations

Data on precipitation rates obtained from meteorological stations (SENAMHI) close to the study locations.

Corresponding study area	Name and location of meteorological stations	Data available
Balsas	Balsas 472501F4: 06°50'39''S, 078°01'50''W, 850 m	September 2007-March 2013 (data missing for some months between October 2009 and November 2012)
Caraz	Yungay 444: 09°08'59.7''S, 077°45'03.7''W, 2527 m	October 2007-September 2012
Tarma	Tarma 554: 11°23'49''S, 075°41'25''W, 3200 m	May 2007-April 2012
Omate	Omate 850: 16°40'39''S, 070°58'57''W, 2080 m	February 2008-January 2013
San Antonio	La Pampilla 839: 16°24'12.2''S, 071°31'00.6''W, 2400 m	October 2007-September 2012
Chuquibamba	Chuquibamba 750: 15°50'17''S, 072°38'55''W, 2832 m; Pampacolca 751: 15°42'51''S, 072°34'03''W, 2950 m; Yanaquihua 864: 15°46'59.8''S, 072°52'57''W, 2815 m	November 2007-October 2012

6.2.3.2 Soil analyses

Soil samples from 3 different locations corresponding to the study areas Caraz, Omate, and San Antonio were collected as representative for the typical soil in the habitat of *K. lappacea*. Soil samples were taken from the top 30 cm (approximately the area where most roots are found), with 5 subsamples from the sites mixed and then analyzed. Soil was not sieved prior to analysis. Soil analysis was performed at the laboratories of the University of Arequipa (UNSA). Soil pH was determined in a ratio 1:1 suspension of soil in water. Organic matter was calculated from organic carbon estimated by oxidization with dichromate in the presence of H2SO4, without application of external heat. Soil texture was determined by the hydrometer method (see Dewis and Freitas, 1984).

6.2.3.3 Population inventories

Population inventories are based on a total of 238 plots (10 x 10 m) at 19 sites in in 6 departments across Andean Peru (see Table 6.1). Plots were delimited in the shape of a square using four ropes (of 10 m length each) with loops at their ends, which were attached to the ground by means of screwdrivers. Adjacent plots were created by leaving two screwdrivers in the ground and moving the two other screwdrivers when delimiting the next plot.

For each plot all individuals were counted and the number of individuals for each of the three size-classes (I) seedling, (II) juvenile, and (III) adult was recorded. (I) Seedlings were defined by the presence of cotyledons only, (II) juveniles by presence of foliage leaves, the low degree of lignification (only main stem lignified), and a maximum of two branching orders, and (III) adults by the degree of lignification (stem and branches lignified), branching at least of three orders, and the formation of generative organs (buds, flowers, fruits). Additionally, excavation holes from recent harvest events were recorded in the study areas in Balsas, Caraz, and San Antonio.

6.2.3.4 Seed material and design of germination experiment

In April 2009, 795 mature *K. lappacea* fruits were collected ca. 4 km NW of Omate (Dept. Moquegua, Peru; 16°39'26.7''S, 071°0'2.76''W, 2500 m) and stored in a zip-loc bag. Six hundred undamaged, fully developed fruits were used for the experiments on germination success, of which 300 were pre-treated as described below. Another seed lot of 257 seeds was collected at the same locality in October 2012.

On October 8th 2011, in total, 6 different experiments were set up with 100 fruits each. Three different sow depths were examined: (I) on the soil surface, (II) in 5 cm, and (III) in 10 cm sow depth. For each of the 3 different sow depth experiments 10 pots (height = 15 cm) were prepared with 10 inserted fruits each. This experimental setting of 3 x 10 pots (3 x 10 pots x 10 fruits = 300 fruits) was prepared duplicatively, i.e., a complete batch of pots was assembled with untreated fruits (as collected from the wild) and a second batch of pots was assembled with pre-treated fruits (thoroughly ground with sand to roughen the pericarp and remove its hairs, Table 6.3).

As substrate an unfertilized mixture of steamed compost and fine river sand (1:1) was used, roughly representing the soil composition and structure of *K. lappacea*'s natural habitat (see Table 6.6). The substrate was mixed all at once in a clean plastic container and then distributed into the pots in order to ensure a homogeneous mixture in the individual set ups. Pots were watered equally only once a week to avoid elevated soil moisture and waterlogging. Temperature range in the greenhouse was 20–27 °C, duration of light was 12 h/day (optimum temperature for germination is reported to be 24 °C for *K. lanceolata*; Musselmann and Mann, 1977).

Seed germination was determined by counting the developed seedlings that were visible at the soil surface and at the close of the experiment (December 15th 2009) by screening the soil in order to separate and count the number of germinated (but at the soil surface not visible) and not germinated seeds. The number of germinated seeds for each treatment was recorded. Host plants are not required for germination, so the experiment was carried out without host plants in the pots (Kuijt, 1969; Musselmann, 1996).

Table 6.3 Germination experiment

Experimental design

Sow depth (cm)	Untreated	Pre-treated
0	10 pots x 10 fruits	10 pots x 10 fruits
5	10 pots x 10 fruits	10 pots x 10 fruits
10	10 pots x 10 fruits	10 pots x 10 fruits

6.2.4 Statistical analyses of data

Statistical analyses were performed using SPSS (Levene's and t-tests; IBM Corp., 2012) and R (Kruskal-Wallis- and Mann-Whitney-U tests; R Development Core Team, 2008).

6.2.4.1 Population inventories

The Levene's Test was applied pairwise among all data sets from the different study sites to test for homogeneity of variance. In order to test for significant differences between the 19 study locations regarding number of counted individuals per size class (seedling, juveniles, adults) and abundance the Kruskal-Wallis-test was applied. Since the data was not normally distributed, the Mann-Whitney-U Test was used post hoc for pairwise testing of the data from the 19 study locations regarding abundance and number of counted individuals per size class.

6.2.4.2 Germination experiment

The significance of differences between number of successfully germinated seeds for each experimental setting was tested pairwise with the help of both Mann-Whitney-U- and t-tests because a part of the data was normally distributed and another part was not.

6.3 Results

6.3.1 Precipitation rates in the study areas

Across all study areas the major portion of rainfall occured in the austral summer leading to a distinctive seasonality: In the study areas situated in the departments Moquegua and Arequipa (i.e., Omate, San Antonio, and Chuquibamba) January and February represented the months with most rainfall, while here all the rest of the year (March–December) monthly rainfalls were extremely low (<20 mm/year) or did not occur at all. This seasonality was less sharply defined in the study areas Balsas, Caraz, and Tarma however, here a period of extremely low monthly precipitation was readily distinguishable between March and December.

Caraz showed by far the highest mean annual precipitation, which was almost double the amount when compared to the area with the next highest quantity of annual rainfalls (Tarma) and more than twice the amount in relation to the average annual precipitation of 335.5 mm/ year across all areas included in the here presented study. Very low annual precipitation rates (>300 mm/year) were found in all other study sites. Five year mean monthly and annual precipitation rates are presented in Table 6.4.

Table 6.4 Precipitation rates in the study areas

[a]Due to some missing data (see Table 6.2) from the meteorological station in Balsas (SENAMHI 472501F4) here a 4 year mean is available only.

Location	5 year mean precipitation rates in the study area 2007–2012(–2013[a]) obtained from SENAMHI												
	Oct	Nov	Dec	Jan	Feb	Mar	Apr	May	Jun	Jul	Aug	Sep	Total/year
Balsas[a]	41.3	55.3	28.2	46.2	22.1	31.5	23.1	12.5	2.6	0.6	2.5	7.4	273.1
Caraz	67.6	91.4	116.4	119.9	107.2	159.1	123.3	13.7	3.9	0.0	0.4	8.9	811.8
Tarma	44.0	34.9	63.7	59.1	72.8	60.2	45.5	11.6	2.4	6.4	4.4	11.0	415.9
Omate	0.0	0.2	19.5	51.8	88.3	12.9	8.8	1.0	0.0	0.0	0.3	0.0	182.9
San Antonio	0.0	0.0	3.5	48.8	59.8	9.8	5.2	0.0	0.0	0.1	0.1	0.0	127.4
Chuquibamba	0.0	0.1	4.7	58.1	107.7	14.8	15.7	0.6	0.0	0.2	0.0	0.1	202.0
Mean	25.5	30.3	39.3	64.0	76.3	48.0	36.9	6.6	1.5	1.2	1.3	4.5	335.5

Table 6.5 Mean monthly and annual precipitation

Data from three meteorological stations (mm) in the region Chuquibamba (October 2009–September 2012).

Period	Oct	Nov	Dec	Jan	Feb	Mar	Apr	May	Jun	Jul	Aug	Sep	Total/year
2009-2010	0.0	0.0	2.1	9.0	39.8	1.9	11.5	2.4	0.0	0.0	0.0	0.0	66.7
2010-2011	0.0	0.0	0.5	22.5	138.6	2.1	14.5	0.8	0.0	0.1	0.0	0.0	179.0
2011-2012	0.0	0.3	16.5	84.4	222.7	45.3	49.7	0.0	0.0	0.0	0.0	0.3	419.2

There is considerable inter-annual variation in overall precipitation. This was particularly pronounced in the study sites in the Departments Arequipa and Moquegua. In the region Chuquibamba (Arequipa), for example, annual mean precipitation was very low in the period October 2009–September 2010 (66.7 mm), almost three times as high in the period October 2009–September 2010 and more than six times as high in the period October 2011–September 2012 (Table 6.5).

6.3.2 Soil analyses

Results from soil analyses are presented in Table 6.6. Soil at the study sites generally was composed of sand, lime, and clay usually without any visible organic matter (1.2 OM% in Caraz). While sand was the dominant soil component (93–94%) at Omate and San Antonio, the portion of lime and clay was notably higher at the sampled site in Caraz. Results from soil analyses were used as a guideline in the experimental design for the experiment on germination success in order to roughly meet the conditions in *K. lappacea*'s natural habitat.

Table 6.6 Results from the soil analyses

Soil composition and pH from three study locations.

Area name	Origin of soil sample	pH	Organic matter (%)	Composition (Sand/Lime/Clay, %)	Textural class
Caraz	Dept. Ancash, Prov. Huaraz, Inner-andean dry valley, 5 km after Caraz on the road to Casma, 08°59'54.1''S, 077°49'24.9''W, 2200 m	7.6	1.2	66.00/20.00/14.00	Sandy loam
Omate	Dept. Moquegua, Prov. General Sánchez Cerro, between Omate and Moquegua, directly beneath Moquegua, Urinay, 16°42'23.3''S, 071°59'40.2''W, 1120 m	6.5	0.0	93.48/4.64/1.88	Sandy soil
San Antonio	Dept. Arequipa, Prov. Arequipa, 3-4km from Yarabamba, 16°34'24.6'S, 071°28'27.36'W, 2545 m	4.4	0.0	94.32/3.32/2.36	Sandy soil

6.3.3 Population inventories

Results from the Kruskal-Wallis-test supported a highly significant difference between the 19 locations in all cases (number of seedlings, juveniles, and adults, and abundance per 100 sqm: $X2 = 136,759$; $118,607$; $131,116$; $143,875$; $df = 18$; $p < 0.001$). The Mann-Whitney-U test was performed for the whole data set in spite of the fact that the Levene's tests revealed that some data sets showed homogeneity of variance while others did not. In the latter case results of the U-test are given in brackets (Appendix B).

Mean abundance across all 19 study locations was 13.7 ± 0.9 *K. lappacea* individuals/100 sqm with by far highest mean values for Caraz (Caraz 2: 54.7 ± 9.2 ind./100 sqm, Caraz 1: 32.1 ± 7.9 ind./100 sqm) and lowest mean values for the plots in San Antonio A1 (0.2 ± 0.2 ind./100 sqm) and Omate (1.6 ± 1.3 ind./100 sqm). Mean number of adults across all study locations was 7.0 ± 0.5 ind./100 sqm with by far the highest number of adult individuals in Caraz 2 (37.7 ± 5.0 ind./100 sqm) and the lowest number in San Antonio A1 (0.2 ± 0.2 ind./100 sqm) and Omate (0.3 ± 0.2 ind./100 sqm, Fig. 6.1; Fig. 6.2 A & B). In spite of the high mean abundances for Chuquibamba 1–4 (26.5 ± 1.6, 28.0 ± 1.6, 24.5 ± 1.2, and 26.1 ± 1.8 ind./100 sqm) the proportion of adult plants was found to be high only in Chuquibamba 3 (17.4 ± 1.1 adult ind./100 sqm) and extremely low in Chuquibamba 2 (3.0 ± 0.5 adult ind./100 sqm), where adult plants made up only 10.7% of the total individuals (compared to the corresponding mean of 51.4% for Chuquibamba 1–4; Fig. 6.1; Fig. 6.2 C & Table 6.7).

Number of juveniles across all study locations was 4.0 ± 0.3 ind./100 sqm with the highest numbers in Chuquibamba 1, 2, and 4 (10.1 ± 1.4, 11.7 ± 1.0, and 11.9 ± 1.1 ind./100 sqm, followed by Caraz 2 (8.4 ± 1.6 ind./100 sqm), and the lowest numbers ascertained in Balsas (1 & 2), Omate (0.4 ± 0.3, 0.7 ± 0.3, and 0.3 ± 0.3 ind./100 sqm), and San Antonio A1 respectively, where no juveniles were present at all. Mean number of seedlings across all study locations was 2.6 ± 0.2 ind./100 sqm with by far the highest numbers in Chuquibamba 1, Chuquibamba 2, and Caraz 1 & 2 (9.6 ± 1.3, 13.3 ± 0.8, 11.3 ± 5.1, and 8.6 ± 3.3 ind./100 sqm) and the lowest numbers in San Antonio B1–3 (0.2 ± 0.1, 0.8 ± 0.4, and 0.6 ± 0.3 ind./100 sqm), Omate (1.0 ± 0.8 ind./100 sqm), and Balsas (1 & 2), Tarma (1 & 2), San Antonio A1, Chuquibamba 5 & 6 respectively, where no seedlings were present at all (Fig. 6.2 A–C & Table 6.7). Mean number of documented digging holes (= recent harvest) ranged from 0.0 ± 0.0 to 5.7 ± 1.2, with the highest numbers recorded in San Antonio B1–3 (Table 6.7). Raw data from population inventories are provided in Appendix B1, and the corresponding results of the statistical analyses are given in Appendix B2.

6.3.4 Fruits and Germination

Of the 795 *K. lappacea* fruits collected in April 2009, only 45 (i.e., ca. 6%) were damaged by rodents or beetles. Conversely, 233 (91%) of the 257 fruits collected at the same locality in October 2012 were damaged and only 24 were undamaged.

Overall range of the documented germination success was 0–10 seeds per pot. Germination success of unburied seeds was significantly lower (28 ± 15% untreated and 16 ± 13% pretreated) than for seeds buried 5 cm (82 ± 9% untreated and 85 ± 7% pre-treated) and 10 cm (88 ± 8 untreated and 86 ± 8 pre-treated, $p < 0.001$ for both t-tests and Mann-Whitney-U tests). No significant differences were found neither between depths of fruit burial (5 vs. 10 cm) nor between treated and untreated seeds. Untreated seeds germinated slightly more

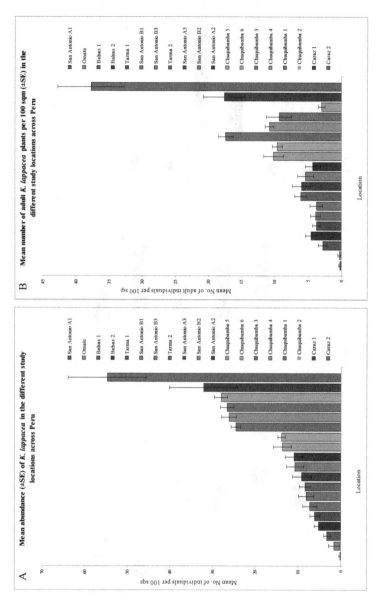

Fig. 6.1 Mean abundance versus mean number of adult *K. lappacea* plants

(A) Mean abundance versus (B) mean number of adult *K. lappacea* plants per 100 sqm (±SE) across Peru.

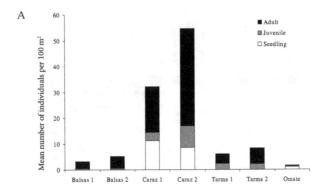

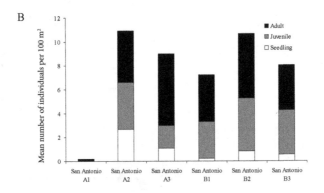

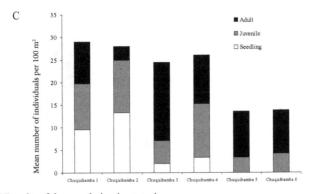

Fig. 6.2 Results of the population inventories

(A–C) Mean number of individuals per size class (seedling, juvenile, adult) and mean total number of *K. lappacea* individuals per 100 sqm (mean abundance) from several study locations across Peru.

Table 6.7 Data from population inventories

Mean number of *K. lappacea* individuals per size class (Seedling, Juvenile, and Adult), mean individuals harvested (digging holes counted) and mean abundance (± SE) from 19 study areas across Peru with (N) number of inventories from each area (10 x 10 m plots). Statistical results are presented in Appendix B2, na – not available.

Study area	N	Seedling	Juvenile	Adult	Harvested	Mean abundance
Balsas 1	16	0.0 ± 0.0	0.4 ± 0.3	2.8 ± 0.7	1.4 ± 0.5	3.3 ± 0.7
Balsas 2	15	0.0 ± 0.0	0.7 ± 0.3	4.5 ± 0.8	0.0 ± 0.0	5.2 ± 1.0
Caraz 1	15	11.3 ± 5.1	3.3 ± 0.8	17.6 ± 3.1	0.0 ± 0.0	32.1 ± 7.9
Caraz 2	7	8.6 ± 3.3	8.4 ± 1.6	37.7 ± 5.0	0.0 ± 0.0	54.7 ± 9.2
Tarma 1	17	0.0 ± 0.0	2.4 ± 0.6	3.8 ± 0.6	na	6.1 ± 1.0
Tarma 2	14	0.0 ± 0.0	2.3 ± 0.6	6.1 ± 0.9	na	8.4 ± 1.3
Omate	10	1.0 ± 0.8	0.3 ± 0.3	0.3 ± 0.2	na	1.6 ± 1.3
San Antonio A1	5	0.0 ± 0.0	0.0 ± 0.0	0.2 ± 0.2	0.0 ± 0.0	0.2 ± 0.2
San Antonio A2	23	2.7 ± 1.1	3.9 ± 0.9	4.3 ± 1.1	2.0 ± 0.4	10.9 ± 2.0
San Antonio A3	12	1.1 ± 0.4	1.9 ± 0.7	6.0 ± 1.3	0.1 ± 0.1	9.1 ± 2.1
San Antonio B1	18	0.2 ± 0.1	3.1 ± 1.0	3.9 ± 0.7	4.6 ± 1.0	7.2 ± 1.6
San Antonio B2	18	0.8 ± 0.4	4.4 ± 1.0	5.4 ± 1.2	4.6 ± 1.1	10.7 ± 2.0
San Antonio B3	18	0.6 ± 0.3	3.7 ± 0.9	3.8 ± 1.0	5.7 ± 1.2	8.1 ± 1.7
Chuquibamba 1	10	9.6 ± 1.3	10.1 ± 1.4	9.4 ± 1.9	na	26.5 ± 1.6
Chuquibamba 2	10	13.3 ± 0.8	11.7 ± 1.0	3.0 ± 0.5	na	28.0 ± 1.6
Chuquibamba 3	10	2.0 ± 0.8	5.1 ± 0.6	17.4 ± 1.1	na	24.5 ± 1.2
Chuquibamba 4	10	3.4 ± 1.0	11.9 ± 1.1	10.8 ± 0.6	na	26.1 ± 1.8
Chuquibamba 5	5	0.0 ± 0.0	3.4 ± 1.2	10.2 ± 1.5	na	13.6 ± 2.1
Chuquibamba 6	5	0.0 ± 0.0	4.2 ± 1.0	9.6 ± 0.7	na	13.8 ± 0.9

successfully on the soil surface than the pre-treated seeds. Results of the germination experiment are presented in Table 6.8, documented raw data is to be found in Appendix B1, and the corresponding results of the statistical analyses (t-test results only) are given in Appendix B2.

Table 6.8 Mean germination success

($\pm$ SE) dependent on sow depth and seed pre-treatment, with N = 10 in all cases. [a], [b] indicate significant differences (p < 0.01) according to both Mann-Whitney-U-tests and t-tests.

Treatment	Untreated			Pre-treated		
Sow depth (cm)	0	5	10	0	5	10
Mean germination success (%)	28.0 ± 4.7^a	82.0 ± 2.9^b	88.0 ± 2.5^b	16.0 ± 4.0^a	85.0 ± 2.2^b	86.0 ± 2.7^b

6.4 Discussion

6.4.1 Abiotic conditions in the study areas

According to the classification system of ecological zones by Jahnke (1982) and the data on annual precipitation rates obtained from SENAMHI (years 2007–2013) all study sites belong to the arid zone, with mean annual amounts of rainfall below 500 mm, except for Caraz, which falls into the semi-arid zone, with a mean annual precipitation of 500–1000 mm (Jahnke, 1982). Overall, our results confirm the findings of a recent study, which reports that K. lappacea is mainly found in areas where mean monthly precipitation roughly ranges between 10 and 75 mm, which corresponds to a mean annual precipitation of 120–900 mm (Giannini et al., 2011).

Regarding soil composition and the proportion of organic matter in the soil, there is a high similarity between Omate and San Antonio, but a notable difference of these two sites to Caraz. This comes about the fact that Omate and San Antonio are located in the Coastal desert and Caraz is situated in an Inner-Andean valley.

6.4.2 Population size and condition

Krameria lappacea populations examined in this study vary notably in size and condition. Clearly, there are several abiotic and biotic factors that impact on their successful regeneration, which represents the key for viable populations to linger: Soil water availability after rainfalls and the resulting soil water potential, e.g., is reported to be one of the main abiotic factors (in addition to temperature) influencing seed germination in plant species that dwell in arid and semi-arid environments (Adams, 1999; Bochet et al., 2007; Flores & Briones, 2001; Noy-Meir, 1973). The presence and survival of seedlings from plant species growing in arid environments has also been reported to be directly linked to water availability (e.g.,

Wilson & Witkowski, 1998). Soil composition decides on the ground's ability and extent to that water may be stored and kept available over time. Thus, precipitation rates, temperature and the resulting evapotranspiration, as well as soil composition determine water availability and by that generally play a significant role in successful seedling recruitment in arid environments, which in turn is essential for a continuous regeneration of plant populations. However, it remains difficult to judge in how far the differences in climate and soil contribute to the differences in abundance, but the data obtained are strikingly parallel to the intensity of harvest. Our results imply that particularly the degree and method of wild harvest represent the most prominent factors pushing the resilience of examined *Ratanhia* populations to its limits.

The populations in San Antonio A1, Omate, and Balsas, where commercial harvest has been intensive – in some cases even marauding – in the past, are in a poor condition, with lowest abundance, lowest number of adult plants and a very low number or complete absence of juvenile stages, which makes continuous existence of these *Krameria* populations uncertain.

Quite the contrary is the case at the study sites close to Caraz. The populations examined here show the overall highest abundance, the highest number of adult individuals, and a high number of juveniles and seedlings, which make out a large proportion of all individuals (1/3–1/2) indicating that these populations are healthy and in a very good condition. Most probably this comes due to a combination of several favorable terms, among which relatively high annual rainfalls and an advantageous soil composition count to the positively affecting abiotic factors. However, even more importantly appears to be the fact that wild harvest is here performed on a low domestic level only, which is confirmed by the absence of the typical digging holes as signs of a recent harvest (see Brokamp *et al.* 2012, see also Table 6.7).

In Tarma and in the remaining plots near San Antonio de Yarabamba (San Antonio A2 & A3, B1–3), overall mean abundance is rather low and does not exceed 11 individuals per 100 sqm, however, here at least one third of the present individuals are in the juvenile stage and accordingly, these populations still are able to regenerate to some degree. All plots near San Antonio are subject to comparable abiotic factors and the difference in size and condition of populations between San Antonio A1 and all the other plots here (San Antonio A2 & A3, B1–3) most probably is due to the fact that the latter are situated within the extractive reserve and thus are better protected than the plot San Antonio A1, which lies just outside the extractive reserve. Although the documented number of up to 18 digging holes per 100 sqm raises the question, whether common harvest levels may be too high to be sustainable even within the extractive reserve.

The plots in the Chuquibamba region represent a special case within our sample and the corresponding results from our investigations shed light on several aspects, which are relevant in the monitoring of plant populations subject to wild harvest in arid environments: The close relation between precipitation and presence of seedlings becomes obvious when relating the number of seedlings documented in these 6 study locations with the corresponding meteorological data at time of data collection. Admittedly, the three meteorological stations in the Chuquibamba region, data was obtained from, are not located at the very same sites and altitudes as the 6 plots. However, the conclusions we draw here are mainly based on the annual and inter-annual seasonality of rainfalls in the region, not on the absolute amounts of local precipitation. In Chuquibamba 5 & 6, where data was obtained in October 2010 (a year with extremely low annual rainfall, dry season, see Table 6.5), not a single seedling was found (see Fig 6.2 C). In contrast to that, the number of present seedlings was low to moderate when annual rainfall was more intensive (Chuquibamba 3 & 4, 2.0–3.4 seedlings/100 sqm, data collected in November 2011, dry season) and highest, when annual rainfall prior to the population inventories was high and data was collected shortly after the rainy season (Chuquibamba 1 & 2: 9.6–13.3 seedlings/100 sqm, data collected in May 2012, cf. Table 6.5 & Fig. 6.2 C). Therefore, seedling numbers alone are a fairly unreliable parameter for determining the condition of populations in arid environments, especially if data are obtained in different years and at the end of the dry season. However, a complete lack of recruitment in Balsas, Tarma, San Antonio A1, and the very low numbers of seedlings in Omate in combination with an overall low abundance and a low number of adults clearly contribute to the impression that overcollection has seriously damaged these populations. Consequently, a better means for assessing the condition of populations provides the number of juvenile individuals and adult plants and the data here show largely congruent results. The low number of adult individuals in Chuquibamba 2 comes due to the fact that this area has just been harvested less than a year before the corresponding field data was acquired. Overall, the *Ratanhia* populations in the Chuquibamba region are in a viable and healthy condition, even given the fact that they are harvested regularly. What makes the difference is that the *Ratanhia* populations here are rather difficult to access, privately protected, and managed in a certain way: Firstly, harvesting is performed in a manner that can be called semi-destructively because commonly only the main root is extracted, while secondary roots are left behind in the soil and often are able to regenerate into individual plants. Secondly, a couple of seeds are reseeded directly into the digging holes, which then are refilled, which fosters continuous recruitment.

6.4.3 Seed dispersal and germination

Seed (fruit) dispersal in *Krameria* is extremely inefficient, with the seeds forming huge coherent mats under the plants, representing several years´ seed production (Weigend, pers. obs.). In theory, fruits should be dispersed by mammals or bird species, since the fruits have hairs with retrorse barbs that catch on feathers, fur, or clothing (Simpson, 2007). Seed burial seems to dramatically affect recruitment: In our experiment we found a threefold increase in germination in buried versus non-buried seed. In nature, the seeds are likely dispersed and buried by rodents (e.g., species of the genera *Abrocoma*, *Lagidium*, *Phyllotis*, among others), which feed on *K. lappacea* seeds and fruits (Holmgren *et al.*, 2001; Kuch *et al.*, 2002; Latorre *et al.*, 2002; R. Aguilar, pers. obs.) and (accidentally or deliberately) collect and store the fruits in rock crevices or their burrows. There is a high probability that the dispersers are – in this case – also the main predators. The fruit lots collected 2009 and 2012 show dramatic differences in the degree of seed predation, which was very low (ca. 6%) in 2009, when *Krameria* was still abundant in the area (>100 individuals on the corresponding slope), and extremely high (91%) in 2012, when only a single shrub of the species was left on the corresponding slope. While these are highly unsatisfactory data, this may indicate that with decreasing population density the effect of seed predation increases exponentially. Seedlings are mostly found directly under mature plants, but if distant from any adult *Krameria* individuals, then they are found in close vicinity to rocks or in channels resulting from rainwater drainage. Thus, wind and water may also play an important role in the secondary dispersal and burial of *Krameria* seeds (compare Simpson, 2007).

6.4.4 Statistical analyses of data from the population inventories

Although in many cases the documented differences regarding number of individuals per size class between several study sites were obvious (e.g., Seedlings Balsas 1 vs. Chuquibamba 2, 0.0 ± 0.0 vs. 13.3 ± 0.8) the statistical means used were not capable of confirming significance due to the immense inhomogeneity of variance within our sample (i.e., through rejection of results from the Mann-Whitney-U tests in case the Levene's test was significant, see Appendix B2).

6.5 Conclusion

Krameria lappacea is perfectly adapted to arid environments and well established in areas where mean annual precipitation ranges between 127 and 812 mm, but which in dry years may receive less than 100 mm. In undisturbed areas that receive relatively high mean amounts of annual rainfall in our sample (i.e., Caraz) the species may form dense populations that comprise more than 60 adult individuals and an overall abundance above 100 individuals per 100 sqm across all age classes.

However, the interaction of abiotic and biotic factors (including anthropogenic influences) put the species under enormous pressure in large parts of its distribution area. Particularly, in years of rigorously high aridity, especially in areas that experience extreme inter-annual variability of precipitation rates the harvest of unsustainable amounts clearly impacts the species´ ability to recover due to an overall very low success of seedling recruitment in these areas. Mismanagement (i.e., destructive harvest and overexploitation) thus easily leads and already has led to depletion of many *Krameria* populations in highly disturbed areas (i.e., Balsas, Tarma, and Omate), rendering this valuable resource commercially extinct. The consequences of dwindling *Ratanhia* populations for its habitat (e.g., increased soil erosion) and the associated flora and fauna (e.g., loss of biodiversity) remain uninvestigated.

In contrast, even in areas that experience very low amounts of rainfall and extreme inter-annual variability of precipitation rates and where *Krameria* roots are harvested regularly (i.e., Chuquibamba and the study areas near San Antonio that are situated within the extractive reserve), it is indeed possible to maintain viable populations for the future by implementation of appropriate management practices. In this, it appears that applied management practices in Chuquibamba (that include the removal of the primary root only while leaving secondary roots in the soil for regeneration and manual reseeding into the digging holes) have an overall lower impact on the populations than the ones applied in San Antonio.

Seed burial is not obligate, but significantly enhances the germination success of *K. lappacea* seeds. In some parts of *K. lappacea's* distributional range the rainfall conditions (i.e., sufficient water ability) required for seed germination occur infrequently only, which leads to a rather episodic recruitment. Especially in years with low rainfall, the implementation of more sustainable management practices, e.g., as performed in Chuquibamba, is strongly recommended in order to foster the conservation of this valuable species.

7 Conclusions

7.1 Trade in palm products in NW South America

7.1.1 Standardized data collection

The ultimate goal was to provide a tool for the collection of significant and interoperable data on production and commercialization of palm resources and products. As published by Johnson (2011) there are two different types of categorization for palm derived products: The first one is based on whether the product is the chief commercial product or whether it is generated during processing or harvesting. The second one is based on type and degree as well as on location and level of sophistication of its processing, which nicely reflects the overall complexity of the topic (see Chapter 1.2.1). The first draft version of the here presented standardized research protocol (SRP, Chapter 2, Appendix A), based on the results from intensive background research on commonly commercialized palm raw materials and products in NW South America (see Chapter 3.3), was consequently quite theoretical and too complex to be handled in the field (Chapter 2.2). Particularly the initial separation into several different questionnaires and corresponding interview forms according to the activity of interviewees, i.e., harvest, transport, (pre)processing and production, or (re)sale, turned out to be counterproductive for achieving the main objectives in designing this protocol (Chapter 2.1). This comes due to the fact that within the value chain of a given palm product individual interviewees may act in more than only one rôle and in addition to that may be actively involved in the production and commercialization of several products and by-products from several species. However, after several revisions and adjustments, as described in Chapter 2.2, and incorporation of additional questions on required permits, official control and problems encountered by stakeholders when pursuing their activity (Appendix A2.1, Master Sheet; H14, H15, H19), the final version is thematically comprehensive, easy to use in the field and allows to obtain a maximum of relevant data in a minimum of time in a standardized manner.

Despite supreme effort to ensure best usefulness of the here presented SRP, its implementation is characterized by inherent limits, which may result in data gaps regarding some of the relevant data from individual interviewees. These limits may either arise on the part of the interviewer, e.g., by inept behavior and failure in gaining the interviewees trust, or on the part of the interviewees for many possible reasons, among which are, e.g., limited self-monitoring, low comprehension of their activities or sometimes limited willingness to share information with outsiders. Thus, certain crucial skills and character traits on behalf of the interviewer are required in order to ensure a successful application of the SRP, such as a basic botanical and commercial knowledge, necessary language skills as well as sufficient empathy and persuasive power. Consequently, a successful standardization of data collection also depends on the performance of individual interviewers, i.e., on the degree to which they understand and comply with the instructions given in the manual (Appendix A1 Manual) and whether data is transferred correctly into the provided data capture table (DCT, Appendix A3). Furthermore, data from verbal reports generally may entail a couple of well known associated problems, such as lack of objectivity, poor memory, or imprecise articulation. Therefore it is recommended to always verify obtained interview data with help of additional information sources (Chapter 2.5; Yin, 2003).

Overall, the here provided SRP, specially designed for the standardized collection of data on trade with palm resources on local to international levels, works well and can be used, e.g., in the data collection prior to a value chain analysis of a given NTFP derived from palms (or from other useful plants). All components of SRP stand by in English and Spanish language and were already in use by colleagues for data collection on trade and commercialization of palm products in Colombia, and Peru (e.g., Vallejo et al., 2011; Vargas Paredes, 2012).

7.1.2 The economically most important palm species

The economically most important palms in northwestern South America are non-native species, such as African oil palm (*Elaeis guineensis* Jacq.) and coconut palm (*Cocos nucifera* L.). All over the tropics both have been cultivated massively and are traded internationally in vast amounts for decades and centuries, respectively. Consequently, both species cover large tracts of cultivated land in South America and lately the African oil palm is increasingly cultivated in plantations in the Andean countries; particularly, in Colombia, but also in Ecuador and Peru (Pacheco, 2012). In Ecuador, e.g., it has been introduced in 1953 and was cultivated since then in order to satisfy the national demand for vegetable oil and to obtain a new product for exportation (Borgtoft Pedersen & Balslev, 1993). Today, Colombia and Ecuador are the main countries producing oil palm in South America with cultivation areas as large as 165,000 and 135,000 ha, respectively (FAOSTAT, 2011). In 2012, Ecuador alone exported 276,000 tonnes of palm oil at a price of around 275 Mio. US$ (Anonymous, 2012). These figures provide insight into the high economic importance and potential of palms in general, indicating the economic significance that single palm species may have, and may act here as a comparative example to economically important native palm species in north western S America.

To date, Colombia, Ecuador, Peru and Bolivia all have a considerable international trade in native palm products. Corresponding data on amounts traded and overall turnover for these export products are documented and available from official trade statistics (Chapter 3.4.1), e.g., regarding Ecuador provided by the Central Bank of Ecuador (in Spanish: Banco Central del Ecuador, BCE). Blunderingly, the formulated export categories used in this context are vague and imprecise, which limits informative content and precision of these data. This is particularly the case in export products where several species provide comparable raw materials that are used in the production of the same or similar products (e.g., palm heart, handicraft products derived from palm timber, fibre or seeds). Internationally traded raw materials or products from different palm species are typically included in wider categories, sometimes even together with non-palm products, which makes relevant data on single palm species imprecise or impossible to ascertain.

Much more problematically, local to national trade in native palm products is only insufficiently controlled and documented by national authorities, if at all, and thus comprehensive data up to the national level are difficult or impossible to ascertain from official statistics in the countries under study (Chapter 3.4.1; De la Torre, 2011).

Despite all that, the data available on international trade clarify that the economically most important product in this context clearly is canned palm heart (*palmito*). Of the four countries under study, Ecuador is the principal exporter of *palmito* in terms of traded amounts and turnover (2012: 31,000 t; 74 Mio. US$; Fig 7.1; BCE, 2012) followed by Bolivia, Peru, and Colombia. Several native species are in use here, either for the domestic or the export market: *Bactris gasipaes* is cultivated for the export (Ecuador, Peru) or domestic market (Colombia), *Euterpe oleracea* and *E. precatoria* are harvested from the wild and exported (Colombia and Bolivia, respectively), and *Prestoea acuminata* (Ecuador) is harvested from the wild and commercialized on a local to regional scale (Chapter 3.3.5; Vallejo et al., 2011).

The second most important native palm product or rather raw material from northwestern South America, which has a long history in international trade (Acosta-Solís, 1948), is vegetable ivory, the hard endosperm of *Phytelephas* species, also known as *tagua* or *corozo*. This raw material is obtained from *Phytelephas macrocarpa* (Peru and Bolivia), *P. aequatorialis* (Ecuador) and *P. seemannii* (Colombia; Chapter 3.3.8; Chapter 4). Vegetable ivory is mainly composed of mannan polysaccharides (70% of mature endosperm; Timell 1957) and is used in the manufacture of a wide array of products. The most important export product from this species represent vegetable ivory discs, called *animelas*, that are mainly exported from Ecuador and used in the production of buttons abroad. Additionally, in all of the four countries there is an extensive handicraft industry producing figurines, jewelry, buttons, etc., which are marketed on a local to international level (Chapter 4.1). For each country export values of commodities produced from vegetable ivory amount to several million US$ per year; exact figures on nationally traded amounts and turnover are unavailable (Chapter 3.4.1). Commonly, mature seeds are collected from the ground in wild stands or management systems that have replaced the original forest, however in times of high demand harvest of immature infructescences occurs, which leads to raw material of inferior quality, called *tagua maceada* (Fig. 4.2 C; Chapter 4.3.4).

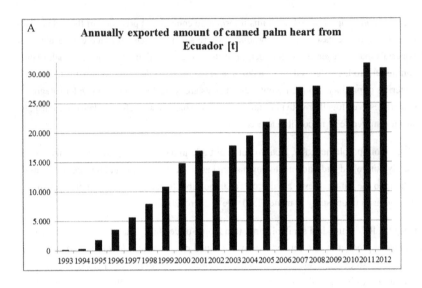

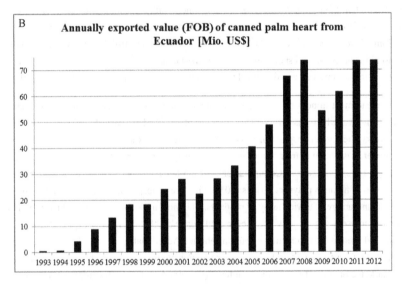

Fig. 7.1 Palm heart exports from Ecuador 1993–2012

(A) Annually exported amount of canned palmito from Ecuador. (B) Total value (free on board) of annually exported canned palmito from Ecuador.

Lack of comprehensive relevant data impedes making further apodictic statements here due to the fact that the remaining palm products are exported in much smaller amounts and trade is difficult to quantify. However, fruit mesocarp (pulp) and seeds (including fruit oil) of species from several genera, such as *Attalea, Bactris, Euterpe, Mauritia,* and *Oenocarpus,* overall are of major significant economic importance or have high economic potential. Native palm fruit is mainly locally commercialized as food and feed, but also regionally to internationally as handicrafts (seeds), and as palm oil, sold pure or as a component in cosmetics (Chapter 3.4.1).

In terms of market volume, however, the fruit of *Mauritia flexuosa,* also known as aguaje, most probably repesents the third most important native palm product, albeit the trade with it is predominantly regional and concentrated on the markets in Iquitos (Peru) and only a tiny fraction (<1 t/a) is exported from here. *Aguaje* forms part of the staple diet in lowland Peru where annually an estimated 1,800–10,000 t are commercialized generating a market value of 0.6–2.5 Mio. US$/a (Chapter 3.36), while markets in Bolivia, Ecuador, and Colombia are comparatively smaller (Castaño *et al.,* 2007; Holm *et al.,* 2008). Harvest is performed destructively by felling (female) trees, which decimates wild stands of this palm since 1990 at an alarming rate (Gilmore *et al.,* 2013). Most of the fruits are boiled and sold as a snack by street vendors in Iquitos (Del Castillo *et al.,* 2006), other final products include *aguaje* soft drinks called *aguajina,* ice cream, popsicles (*chupetes*), or frozen pulp, all of which are traded locally to nationally (Kahn, 1991; Navarro, 2006; Delgado *et al.,* 2007; Chapter 3.4.1). Next in line here may be fruits and derived products from *Attalea*-species, *Bactris gasipaes, Euterpe*-species or *Oenocarpus bataua,* which are commercialized locally to regionally either unprocessed or as ingredients for food and cosmetic products (Chapter 3.36).

The commercialization of products based on leaf (*Astrocaryum* species) or leaf sheath fibre (*Leopoldinia piassaba, Aphandra natalia;* Chapter 3.3.4) and of leaves as ceremonials (*Ceroxylon,* Chapter 3.3.3) or thatch (*Lepidocaryum tenue, Geonoma deversa, Phytelepha*s-species, among others, Chapter 3.3.2) can be important drivers of local or regional economies and often represent one of the most important sources of cash income for many indigenous families and communities in northwestern South America (Chapter 3.4.1). All species mentioned here provide products that are derived from palm leaves and probably belong to the economically most important palms in that context, however they represent only a fraction of palm species in use here, since 111 palm species in 37 genera were recorded as sources of fiber in South America. Further, it is reported that a number of species are only mentioned as a fiber source in ethnobotanical lists, but in depth research about their use has not been done yet, in spite of field observations suggesting that their use is extensive and important (Isaza *et al.,* 2013).

The most important palm species in the locally to regionally proceeding trade with palm timber and derived products probably represents *Iriartea deltoidea*, albeit many other species are also in use, particularly where *Iriartea* is less present (Montúfar and Pintaud, 2006), among which are *Socratea exorrhiza*, *Bactris gasipaes*, *Wettinia-* and *Ceroxylon-* species. The product range includes locally sold stem splits and laths used for the construction of fences and houses as well as regionally sold lats used in the construction of supporting structures for the cultivation of bananas and cut flowers. Other more elaborate palm timber products are also commercialized internationally, such as parquete flooring, furniture, utensils, and souvenirs (Chapter 3.3.1; 3.4.1).

As individual economically important palm species may provide several raw materials or products (e.g., vegetable ivory and thatch from *Phytelephas*-species, see Chapter 4.1; palm heart, fruit, and timber from *Bactris gasipaes*; palm heart, fruit, and seeds from *Euterpe*-species) and part of relevant data is unavailable, it remains difficult to assess the overall economic importance (including all products provided) of these species. In this context it is also of importance that the simultaneous exploitation of different plant parts from the same palm species (and palm stands) can be a limiting factor for one or all of the associated economic activities based on it. The case study on *Phytelephas aequatorialis*, e.g., revealed that harvest of leaves negatively affects fruit production (Chapter 4.4.4). For other palm species that provide several raw materials, however, insightful data in this context remains largely unavailable.

7.1.3 Socio-economic importance

There is much literature on palm use in tropical America that provides insights into the socio-economic impact of palm products (e.g., Lévi-Strauss ,1952; Macía, 2004; Paniagua-Zambrana *et al.*, 2007; Macia *et al.* 2011). Non-timber forest products, including many palm products, are generally accepted as important sources of income for rural dwellers (Stoian, 2005; Sunderlin *et al.*, 2005), but quantitative information on the role these NTFPs are playing in local economies is virtually non-existent (Padoch, 1987; Pinedo-Vasquez *et al.*, 1990).

There are only few comprehensive studies on current trade volumes, economic potential, and value chains of palm products. Most of these studies focus on single products offering differently detailed information on few palm products from one or few communities only, which leads to incomprehensive data in this context.

However, it is necessary to understand the value chains for individual palm products in order to reveal the share of benefits obtained by stakeholders and it is particularly important to disclose how individual palm species are valued by primary producers (Chapter 1.3.2; Newcome *et al.*, 2005). Detailed information on structure of value chains, volumes traded, and economic potential also represent a crucial basis for the development of current and future markets (Chapter 3.1) as well as for assessing the value of single species and the associated ecosystems (Chapter 1.1 & 1.3.2; Newcome *et al.*, 2005).

In northwestern South America palm resources represent an important component in day-to-day life of numerous rural communities and the bulk of palm raw materials directly provides them with some of the basic necessities for subsistence, i.e., fruit as (staple) food and leaves and stems as material for the construction of houses, and palm raw materials or products are bartered for other goods, such as clothing or fuel (= immediate use sensu Johnson, 2011). Another large proportion is marketed directly, i.e., the family that harvests the product sells it directly to the consumer at the local market. If processing is involved (boiling of fruits, manufacture of handicraft, small scale processing into oils, etc.; = cottage-level processing sensu Johnson, 2011), then this will typically be carried out by family members. In this case (the family of) the primary producer receives the added value (gross benefit). For handicraft sold at the national level, where value chains still are short and simple, a relatively high proportion of the retail price percolates back to the primary producers (*Astrocaryum chambira* carrying bags, Ecuador: 50–60%; *Astrocaryum chambira* hammocks, Ecuador: 30–60%; thatch, Peru: 55%), which often represents more than 30% of overall monetary income for individual households (Chapter 3.4.1). Regarding palm products that come from small scale industrial processing (sensu Johnson, 2011) the primary producer's proportion of the retail price, paid by the domestic industry for raw material sourcing, is still considerable (e.g., 12%, *Oenocarpus bataua* fruit for oil extraction, Bolivia). However, with increasing geographical extent, trade volumes, and industrialization, the costs of raw material in relation to the retail price of finished products becomes increasingly reduced. This is the case for products such as canned palm heart (Bolivia) and furniture from *Iriartea* timber (Ecuador); here the corresponding figures are only 2–6%. In Peru, the price fetched by the primary producer of the unprocessed vegetable ivory nuts is only 1.6% of its export price (FOB) and 0.1% of the price for nuts sold on the German retail market. Most of the profit from export products is thus typically generated further up in the value chain, and often abroad (Chapter 3.4.2). Overall it is clear that, several palm products (*Astrocaryum chambira* fibre, *Oenocarpus bataua* and *Mauritia flexuosa* fruit, *Lepidocaryum* leaves, *Phytelephas* nuts, etc.) provide the only or most important sources of cash income for numerous local communities and are crucial for local – and sometimes regional – economies. Trade volumes, as measured in metric tons and US$ and as captured in official statistics, do not adequately reflect the tangible socio-economic value of these products and grossly underestimate their socio-economic importance.

Value chains in palm products are heterogenous, strongly depending on product type, market penetration and the number of middle men involved. Consequently, value chains may differ notably even within the same region and for the same product (Chapter 3.4.2).

7.1.4 Sustainability

Colombia, Ecuador, Peru and Bolivia have all implemented legislation for prohibiting "non-sustainable" use of forest resources (Chapter 1.3.4; De la Torre *et al.*, 2011). There is, however, just little evidence that these laws have a noticeable effect on how resources are managed. To date, sustainable harvest of palms is a rare exception (Bennett 2002), although the harvest for only few palm products necessarily needs to be destructive, namely palm timber and palm heart. The bulk of trade in native palm products in north-western South America is still largely based on destructive harvest of wild populations. Unnecessary felling to harvest palm leaves or fruits is a widespread practice, overharvest is common, and mismanagement prevails (Balslev, 2011; Bernal *et al.*, 2011). Therefore, and due to the fact that the bulk of trade in palm raw materials is informal, legislation is largely ineffective. There are, nevertheless, some positive examples, for effective legal protection and largely sustainable use (*Ceroxylon* leaves, *Astrocaryum* fibre, *Phytelephas* vegetable ivory, *Aphandra fibre, palmito* from cultivated *Bactris gasipaes*, *Lepidocaryum* leaves for thatch (Chapter 3.4.3; Bernal *et al.*, 2011; Galeano & Bernal, 2010; Borgtoft Pedersen & Balslev, 1993). In *Phytelephas* species, comparatively lower benefits in the trade with *tagua* nuts (collection of fruits) support a simultaneously proceeding trade with thatch (leaf harvest), which together with the habit of harvesting immature infructescences (*tagua maceada*) negatively affects resource availabilty, resource quality, and may impair its sustainable use (Chapter 4.4.4). To date, still several tens of million palm trees are cut down annually and, for the main part, unnecessarily to obtain fruits or leaves. Rising prices and reduced availability of raw materials is already widespread in a range of products. Much of the palm extraction is therefore based on shifting extraction practices where new areas are harvested every year and the resources are already depleted near villages and urban areas (Chapter 3.4.3; Kvist & Nebel, 2001).

Most NTFP markets are small in scope and value, and therefore attract limited attention or investment only. But when they do become successful, sustaining supply may be a serious problem (Shanley *et al.*, 2002). Especielly, unless destructive harvest is replaced by non-destructive practices, which are available for almost all of the species, even the most common palms such as *Mauritia flexuosa, Oeonocarpus bataua* and *Euterpe* spp. will face commercial extinction. Since destructive harvest is often selective, the best-quality trees are lost from the gene pool. As a consequence, genetic resources rapidly degrade and reduce the options for future resource development. Figures concerning the relative abundance of destructive harvest as opposed to sustainable harvest for other major palm products are not available, but are urgently needed for adequate resource management. There is a growing understanding of the negative impacts of destructive harvest and progress has been made in several regions towards implementing more sustainable harvest techniques. Overall, the development, establishment and control of sustainable harvest techniques are probably the most important requirements for positive mid-term developments. Possible threats against this scenario are bad governance, corruption, and uncertainty of land tenure. Serious impediments in both policy making, law enforcement, and research will have to be overcome to develop the full potential on an ecologically and economically sustainable basis (Chapter 3.4.3). Also, the value chains will need to be adjusted, so as to ensure that primary producers receive an adequate share of the overall benefits. This has to go hand in hand with strict policing and the clarification of contentious issues such as land tenure and ownership of the resources -otherwise higher profits will be an incentive for the informal and destructive harvest (Chapter 3.4.4).

7.1.5 Productivity and management of *Phytelephas aequatorialis*

7.1.5.1 Leaf and fruit production

There is major variation in leaf production of *P. aequatorialis* (4.1–13.3 leaves per year), depending on location characteristics and gender of the individuals: Leaf production is positively correlated to exposure to sun light, lowland palms produce more leaves than highland palms, and male individuals produce more leaves than females. In order to harvest a maximum number of leaves with minimum effort, individual male palms thus should be harvested approximately every two years, exposed palms slightly more often than shaded ones. In areas of sourcing for vegetable ivory the bulk of female palms should not be harvested for leaves at all.

Fruit development is extremely slow in *P. aequatorialis*, ranging between three and eight years with longest development times in the highlands. Annual infructescence production per (female) individual ranged immensely (0.8–6.7 infructescences with ca. 120 fruits each) and in the lowlands is at least twice as high than in the highlands. A conservative estimate of the annual production of one hectare with 500 (250 female) *Phytelephas* individuals in the highlands spans from 750–2,500 kg vegetable ivory nuts (dry weight), however, annual yields appear to be considerably higher in the lowland. A correlation between exposure to light and fruit production was found in lowland agroforestry, but this was not the case for other land use systems within our sample.

7.1.5.2 Impact of leaf harvest on fruit production

Today, in coastal Ecuador predominantly tourist facilities are constructed in traditional architecture which includes palm thatch roofing. Therefore, there is a small but notable local market based on leaves for thatch in touristic regions of western Ecuador. Due to high prices, a short value chain, and considerable demand for *Phytelephas* thatch, the commercialization of leaves, hence, sometimes represents a more lucrative business for farmers in these areas. Shift from commercialization of vegetable ivory nuts to leaves for thatch may result in unsatisfied demand and rising prices of raw material in the national *tagua* industry and may further lead to resource limitation for companies exporting vegetable ivory discs (animelas). Measures should be taken to ensure sustainable use and commercialization of the two partially exclusive and locally competing products because leaf harvest considerably reduces vegetable ivory production, even when enough leaves are left on female palms.

7.2 Biology, management and sustainability of *Krameria lappacea*

7.2.1 Parasitism, seed ecology, abundance, and harvest impact

Krameria lappacea is perfectly adapted to arid environments and well established in areas where mean annual precipitation ranges between 127 and 812 mm, but which in dry years may receive less than 100 mm. In undisturbed areas that receive relatively high mean amounts of annual rainfall the species may form dense populations that comprise more than 60 adult individuals and an overall abundance above 100 individuals per 100 sqm across all age classes. *Krameria lappacea* populations examined in this study varied notably in size and condition. Clearly, there are several abiotic and biotic factors that impact on their successful regeneration, which represents the key for viable populations to linger. However, it remains difficult to judge in how far the differences in climate and soil contribute to the differences in abundance, but the data obtained are strikingly parallel to the intensity of harvest.

Krameria lappacea is indeed terrestrial hemiparasitic forming haustorial connections to several host plant taxa. The haustoria of *Krameria lappacea* usually are rounded, blackish structures formed on roots of its hosts. Both fine roots and root hairs are absent in young plants which reflects the obligate hemiparasitism of the species. Haustorial connections of the root systems of *Krameria* were found to a wide range of host plants of 18 different species from 17 genera in 12 different plant families including herbaceous and woody taxa and angiosperms and gymnosperms (Chapter 5.3). *K. lappacea* shows rather primitive characters for a terrestrial hemiparasite in terms of specialization, such as the exclusive formation of secondary haustoria, lack of organ reduction (stems and leaves normally developed), large seeds, low host-specificity and the ability to germinate in the absence of a host Weber (1993).

The zygomorphic oil-flowers develop into large, one-seeded, glochidiate fruits typical of the genus (Simpson, 1982, 2007). In theory, fruits should be dispersed by mammals or bird species, since they have hairs with retrorse barbs that catch on feathers, fur, or clothing (Simpson, 2007). In nature, the seeds are likely dispersed and buried by rodents, which collect *K. lappacea* fruits and feed on them or store the seeds in rock crevices or their burrows. There is the possibility that here the dispersers also act as main predators. However, seed (fruit) dispersal in *Krameria* can be extremely inefficient, with the seeds forming huge coherent mats under the plants, representing several years' seed production (Weigend, *pers. obs.*).

Seed burial seems to dramatically affect recruitment: In our experiment we found a threefold increase in germination in buried versus non-buried seed.

Our results imply that particularly the degree and method of wild harvest represent the most prominent factors pushing the resilience of examined *Ratanhia* populations to its limits and clearly determines the degree of disturbance and impact in these ecosystems. Data indicates that with decreasing population density the effect of seed predation increases extremely. Due to meagre available data a deeper understanding of the ecological role *Krameria* plays in its ecosystem still lacks. However, the implementation of more sustainable management techniques should have high priority, which include (I) establishment of harvest rates adjusted to the condition of individual populations, (II) lower harvest impact by cutting primary roots while leaving secondaries to regenerate, (III) reseeding and refilling of digging holes to foster continuous recruitment (Chapter 6.4.3) Cultivation of *Krameria* to produce the *Rhatany*-roots for the national and international trade will be complicated by the need to cultivate it together with at least one of its host plants. On the other hand, it seems unlikely that the availability of host plants is a limiting factor for the natural distribution and abundance of *K. lappacea*, since it appears to be very unspecific in its host plant choice (Chapter 5.4).

Recent studies clearly indicate that parasitic plants may play a crucial role in maintaining biodiversity (Callaway and Pennings, 1998; Joshi *et al.*, 2000; Press and Phoenix, 2005; Pennings and Callaway, 1996; Watson, 2009). This may be particularly true for *K. lappacea* in its (semi-) desert habitats, where a relatively small number of other perennial species are present and likely all of them are affected by *K. lappacea* in their ecological performance, possible leading to reduced vigour and cover, as shown for other root parasites (Joshi *et al.*, 2000; Chapter 5.4).

Abstract

Biology, sustainability and socio-economic impact of wild plant collection in NW South America

All over the world and especially in the northwest of South America wild plants and plant resources are of high ecological, cultural and economic - often substantial – importance for countless people. Particularly in the species-rich tropics a wide variety of different plants provides an enormous range of essential ecosystem services that are used directly or indirectly by humans. Indirect use here refers to ecosystem services (such as, biomass and oxygen production by photosynthesis, carbon dioxide sequestration, water purification, erosion control, etc.), direct use represents the collection and use of ecosystem goods, i.e., the harvesting and utilization of plant resources (such as timber, fruits, leaves, roots, seeds, fibres, resins, waxes, etc.).

This work focuses on two main subject areas in this context, which are presented separately in the different chapters: After a general, cross-thematic and contextualizing introduction (Chapter 1), the following three chapters deal with questions about the scope, sustainability and economic importance of traded resources from native palms (Arecaceae) in northwestern South America (Chapter 2-4). The second part of this thesis deals with botanical and ecological aspects and the sustainability in the management of wild stocks of a medicinal plant from the Peruvian Andes (*Krameria lappacea*) and the impact of wild harvest on the condition of individual populations (Chapter 5 & 6).

The two topics are processed separately, with each chapter corresponding to a part of the work that has already been published or is in preparation to be published in scientific journals (see page 26 & 27). Each chapter therefore includes separate sections for introduction, materials and methods, results, and discussion. Chapter 7 summarizes the results of Chapters 2-4 (Arecaceae) and Chapter 5 & 6 (*Krameria lappacea*) and forms the conclusion of this study, which was based on the following research questions:

1. How can the data on trade with wild plant resources be obtained in a standardized manner?

2. What are the economically most important native palm species, raw materials and products in northwestern South America in terms of turnover and amounts traded?

3. What is the overall socio-economic importance of the trade in palm resources for primary producers and which share of the overall benefits do they obtain?

4. How are leaf and fruit production in *Phytelephas aequatorialis* correlated to environmental factors (altitude and exposure to sunlight) versus management?

135

5. What is the biology of *Krameria lappacea* and how can sustainable management be based on a biological understanding of the species?

To answer these questions both botanical methods were applied in field and laboratory work and in experiments in the greenhouse and botanical economic data were analyzed.

Chapter 1 presents the here treated plants, explains the relationship between ecosystem services and the use of wild plants, reports on the legal backgrounds of wild harvesting and marketing of vegetable raw materials in South America and presents a typical analysis tool of economic botany. Furthermore, the objectives of the work are listed and the research approach is explained and a general overview of the structure of the dissertation is given.

Chapters 2-4 deal with questions about the scope, sustainability and economic importance of the trade resources with native palm (Arecaceae) in northwestern South America.

In Chapter 2, a standardized research protocol (SRP) is presented, which was designed as a simple and universally applicable tool for capture of relevant data on the production and commercialization of palm products by means of standardized surveys of stakeholders. The SRP consists of three components: (I) the manual with instructions for conducting interviews, including a questionnaire, (II) the forms for data recording during the interviews and (III) the data capture table (DCT) for the transfer and consolidation of data. The questionnaire includes all necessary questions, the answering of which represent relevant data that are necessary for the understanding of the detailed processes of all commercial activities (e.g., questions about origin, quantity and type of raw materials used, the timing and nature of manufacturing processes; expenses and profits of individual stakeholders, trade limiting factors). General problems and limitations in the application of the SRPs are discussed.

Chapter 3 presents the results of an extensive revision on trade in palm products in northwestern South America, which is based on over 200 scientific publications and internet sources. Palm trees are particularly present near the equator and are known worldwide as third most important crop family after the grasses and legumes, as they provide a variety of raw materials and products. Besides wood, countless non-timber products (NTFPs) are used and marketed. The bulk of palm products is used directly and has particularly in rural communities of tropical countries many important applications in subsistence. Palms provide a wide range of resources ranging from food and feed, construction materials and raw materials for traditional or modern crafts through to cosmetics and medical applications. Trade with palm products takes place at the local, regional, national and international level. This work is focused on value chains, volumes traded, prices, and the recent developments of some of the most economically important, native palms and the raw materials they provide. Socio-economic aspects are also presented and discussed. For many households, the marketing of palm resources or products represents one of the most important options to participate in the cash economy, which is essential to provide access to health care and education. Overall, native palm products are playing an important socio-economic role in local, regional and sometimes national markets ("drivers of economy") in South America. However, the number of traded native palm products decreases with increasing geographical extent, i.e., the number of products that are traded internationally is relatively small and represents only a fraction of the usable diversity of resources. Nevertheless, some native palms are of particular economic importance since they deliver export products with turnovers reaching tens of million US$, such as palm hearts and vegetable ivory.

Chapter 4 is a case study on productivity and sustainable management of *Phytelephas aequatorialis*, a palm endemic to western Ecuador. This species provides vegetable ivory (tagua, the endosperm) for handicraft and also leaves for thatch. Harvest of commercialized resources is mainly from nature, although the palm is occasionally cultivated. Most seeds are collected from the ground. In times of high demand, however, some primary producers tend to collect immature inflorescence; these young seeds are of low quality and not suitable for the production of tagua products for export. The development of the inflorescence takes at least three years in the lowlands and over four years at the Andean slopes at around 1,400 m above sea level. Data from 365 palms included in this study showed that male palm trees produce significantly more leaves than female palms. The harvest of leaves has little effect on the production of leaves, but the fruit production is significantly reduced. Sustainable use and marketing of the two partially exclusive and locally competing products tagua (vegetable ivory) and cade (leaves for roofs) have to be carefully designed. Application of non-sustainable practices in the harvest of seeds and leaves, moderately efficient regeneration of populations in pastures and insufficient availability of high-quality raw material for the tagua manufacturing industry represent the greatest challenges in the sustainable utilization of this valuable palm species in the future.

Chapter 5 and 6 deal with botanical and ecological aspects and backgrounds of sustainability in the management of wild stocks of a medicinal plant from the Peruvian Andes (*Krameria lappacea*) and the impact of wild harvest in the state of individual populations.

In Chapter 5 the results of the study of parasitism and haustorium anatomy of *Krameria lappacea* (*rhatany*, Krameriaceae) are presented. *Rhatany* is an endangered medicinal plant from the semi-desert of the South American Andes and is destructively harvested from nature. The present study investigated the presence or absence of hemiparasitism, the host plant spectrum and morphology and anatomy of the haustoria. Hemiparasitism could be confirmed and 106 haustorial connections to 18 host plant species from 17 genera and 12 plant families were recorded. By the results of this study, the number of known host plants was doubled for *Krameria* and the list of hosts was extended by four angiosperm families. Overall, *K. lappacea* is a very generalistic hemiparasite, which probably influences the performance of most perennial species in its habitat. Over-collection therefore probably affects the entire vegetation notably. Strategies for the conservation and sustainable harvesting should have high priority due to the special ecological and economic importance of this species.

Chapter 6 deals with the effects of wild harvest on the condition of natural *Krameria lappacea* populations and the question how to make the sustainable wild harvesting based on a better understanding of the biology of the species. *Krameria* root extracts are used as a promising ingredient for various cosmetic and pharmaceutical preparations. However, commercial cultivation of this medicinal plant is not economical. The commercial sourcing takes place mainly in Peru. By destructive harvesting techniques and extensive marketing wild stocks of this type are becoming increasingly rare, and in some areas *rhatany* populations are already heavily degraded (commercially extinct). The study is based on abundance data from the 238 examined plots (100 square meters) in 6 regions in Peru and the conditions of the differently affected populations are compared and discussed. Experimental results on the germination of *rhatany* are presented, which have relevance for available management options. Strategies for the conservation of *K. lappacea* should have high priority and it is strongly recommended to promote the implementation of sustainable harvesting practices.

In Chapter 7, the major findings of each chapter are finally summarized and discussed separately for both main subjects of this work.

Zusammenfassung

Biology, sustainability and socio-economic impact of wild plant collection in NW South America

Auf der ganzen Welt und besonders im Nordwesten Südamerikas sind wilde Pflanzen und Pflanzenressourcen für unzählige Menschen von hoher ökologischer, kultureller und wirtschaftlicher - häufig substanzieller - Bedeutung. Besonders in den artenreichen Tropen liefert eine Vielzahl von unterschiedlichen Pflanzen eine enorme Bandbreite an essentiellen Ökosystemischen Leistungen, die direkt oder indirekt vom Menschen genutzt werden. Indirekte Nutzung bezieht sich hier auf Ökosystemische Leistungen (wie z.b., Biomasse- und Sauerstoffproduktion durch Photosynthese, Kohlendioxid-Sequestrierung, Wasserreinigung, Erosionsschutz, etc.), direkte Nutzung meint Entnahme und Gebrauch von Ökosystemischen Gütern, d.h. die Ernte und Nutzung von pflanzlichen Rohstoffen (wie Holz, Früchte, Blätter, Wurzeln, Samen, Fasern, Harze, Wachse, etc:).

Im Zentrum dieser Arbeit stehen in diesem Zusammenhang zwei Themenbereiche, unterteilt in mehrere Teilaspekte, die separat in einzelnen Kapiteln bearbeitet werden. Nach einer allgemeinen, themenübergreifenden und kontextualisierenden Einleitung (Kapitel 1) beschäftigen sich die folgenden drei Kapitel mit Fragen zu Umfang, Nachhaltigkeit und wirtschaftlicher Bedeutung des Handels mit nativen Palmenressourcen (Arecaceae) im Nordwesten Südamerikas (Kapitel 2-4). Der zweite Teil der Arbeit befaßt sich mit botanischen und ökologischen Aspekten und Hintergründen zur Nachhaltigkeit in der Bewirtschaftung wilder Bestände einer Medizinalpflanze aus den peruanischen Anden (*Krameria lappacea*) und den Auswirkungen der Wildernte auf den Zustand einzelner Populationen (Kapitel 5- 6).

Die beiden Themenbereiche werden separat bearbeitet, wobei die einzelnen Kapitel Beiträgen entsprechen, die bereits in Fachzeitschriften veröffentlicht wurden, bei diesen eingereicht sind oder in Kürze eingereicht werden. Dies ist den jeweiligen Fußnoten auf Seite 26 und 27 zu entnehmen. Jedes Kapitel enthält dementsprechend eigene Abschnitte zur Einleitung, zu den Materialien und Methoden, Ergebnissen und eine Diskussion. Kapitel 7 diskutiert zusammenfassend die Ergebnisse der Kapitel 2-4 (Arecaceae) bzw. Kapitel 5-6 (*Krameria lappacea*) und bildet den Abschluß der vorliegenden Arbeit, der folgende wissenschaftliche Fragestellungen zu Grunde lagen:

1. Wie können Daten über den Handel mit pflanzlichen Ressourcen aus Wildernte in standardisierter Form aufgenommen werden?

2. Welche sind die wirtschaftlich bedeutendsten einheimischen Palmenarten, Rohstoffe und Produkte in Bezug auf Umsatz und Handelsvolumen im nordwestlichen Südamerika?

3. Welche sozioökonomische Bedeutung hat der Handel mit Palmrohstoffen für Primärproduzenten und wie hoch ist ihre Profitbeteiligung?

4. Welchen Einfluß haben Umweltfaktoren (Höhenlage und Lichtintensität) und Bewirtschaf-
tungweise auf Blatt- und Fruchtproduktion bei *Phytelephas aequatorialis*?

5. Was ist die Biologie von *Krameria lappacea* und wie kann nachhaltiges Management auf
dem biologischen Verständnis der Art basieren?

Um diese Fragen zu beantworten wurden sowohl botanische Methoden in Feld- und Labo-
rarbeiten angewandt, als auch Experimente im Gewächshaus durchgeführt und ökonomisch
botanische Daten analysiert.

Kapitel 1 stellt einleitend die hier behandelten Pflanzen vor, führt in die Thematik ein und
erläutert den Zusammenhang zwischen Ökosystemischen Leistungen und der Nutzung von
Wildpflanzen, berichtet über die legalen Hintergründe von Wildsammlung und Vermarktung
pflanzlicher Rohstoffe in Südamerika und stellt ein in der ökonomischen Botanik eingesetz-
test Analysewerkzeug vor. Weiterhin werden die Ziele der Arbeit aufgeführt und es wird
erläutert, welche Ansätze zur Bearbeitung der Themenkomplexe gewählt wurden und eine
Übersicht über die Gliederung der Arbeit gegeben.

Kapitel 2-4 beschäftigen sich mit Fragen zu Umfang, Nachhaltigkeit und wirtschaftlicher Bedeutung des Handels mit nativen Palmenressourcen (Arecaceae) im Nordwesten Südamerikas.

In Kapitel 2 wird ein Standardisiertes Forschungsprotokoll (SRP) vorgestellt, welches
speziell als einfaches und universell anwendbares Werkzeug entwickelt wurde, um relevante
Daten zur Produktion und Vermarktung von Palmenprodukten mit Hilfe von Befragungen
Beteiligter in einheitlicher Form aufzunehmen. Das SRP besteht aus drei Komponenten: (I)
Anleitung zur Durchführung von Interviews und inklusive Fragenkatalog, (II) Formulare für
die Datenaufnahme während der Befragungen und (III) Datenaufnahmetabelle (DCT) zur
Übertragung und Zusammenführung von Daten. Der Fragenkatalog bezieht alle notwendi-
gen Fragen mit ein, deren Beantwortung relevante Daten liefern, die für das Verstehen der
genauen Abläufe aller kommerziellen Aktivitäten notwendig sind (z.B., Fragen zu Herkunft,
Menge und Typ verwendeter Rohstoffe; Art und Ablauf von Verarbeitungsprozessen; Aus-
gaben und Gewinne einzelner Beteiligter; Faktoren, die den Handel limiteren). Generelle
Probleme und Limitierungen in der Anwendungen werden diskutiert.

Kapitel 3 stellt die Ergebnisse einer umfangreichen Revision über den Handel mit Palmenprodukten im nordwestlichen Südamerika vor, die auf über 200 wissenschaftlichen Publika-
tionen und Internetquellen basiert. Palmen sind in Äquatornähe besonders präsent und gelten

weltweit als drittwichtigste Nutzpflanzenfamilie nach den Süßgräsern und Hülsenfrüchten, da sie eine Vielzahl von Rohstoffen und Produkten liefern. Neben Holz werden auch zahllose Nicht-Holzprodukte (NTFPs) genutzt und vermarktet. Der größte Teil der Palmprodukte wird direkt genutzt und findet besonders in ländlichen Gemeinden tropischer Länder vielerlei subsistenziell bedeutende Anwendungen. Angefangen bei Nahrungsmitteln und Futtermitteln über Baumaterialien und Rohstoffe für traditionelles oder modernes Handwerk bis hin zu Kosmetika und medizinischen Anwendungen stellen Palmen eine große Bandbreite an Ressourcen bereit. Handel mit Palmprodukten erfolgt auf lokaler, regionaler, nationaler und internationaler Ebene. Der Fokus dieser Arbeit liegt auf Wertschöpfungsketten, Handelsvolumen, Preisen, und den jüngsten Entwicklungen für einige der ökonomisch bedeutendsten, heimischen Palmen und deren Rohstoffe. Sozioökonomische Aspekte werden ebenfalls vorgestellt und diskutiert. Für viele Haushalte stellt die Vermarktung von Palmressourcen oder -produkten eine der wichtigsten Möglichkeiten dar auch über den Tauschhandel hinaus an der Geldwirtschaft teilzunehmen, was u.a. entscheidend für den Zugang zu ärztlicher Versorgung und Bildung ist. Insgesamt spielen native Palmprodukte in Südamerika also eine wichtige sozioökonomische Rolle in lokalen, regionalen und manchmal nationalen Märkten („drivers of economy"). Die Zahl der gehandelten nativen Palmprodukte nimmt jedoch mit steigendem geografischen Ausmaß deutlich ab, d.h., die Zahl der Produkte, die international gehandelt werden ist vergleichsweise gering und stellt nur einen Bruchteil der nutzbaren Ressourcenvielfalt dar. Native Palmen von besonderer ökonomischer Bedeutung liefern Exportprodukte mit Umsätzen in zweistelliger Millionenhöhe, wie z.B. Palmherzen und pflanzliches Elfenbein.

Kapitel 4 ist eine Fallstudie zu Fragen der Produktivität und der nachhaltigen Bewirtschaftung von *Phytelephas aequatorialis*, eine im Westen Ecuadors endemisch vorkommende Palme. Von dieser Art werden pflanzliches Elfenbein (*tagua*, das Endosperm) und Blätter für das Decken von Palmdächern vor allem aus der Natur geerntet und kommerzialisiert, obwohl die Palme gelegentlich angebaut wird. Die meisten Samen werden vom Boden gesammelt. In Zeiten hoher Nachfrage jedoch, sammeln einige Primärproduzenten unreife Fruchtstände; diese jungen Samen sind von minderer Qualität und nicht geegnet für die Herstellung von tagua Produkten für den Export. Die Entwicklung der Fruchtstände dauert im Tiefland drei Jahre und über vier Jahre an den Anden-Hängen bei rund 1400 m ü.N.N. Daten von 365 untersuchten Palmen zeigen, dass männliche Palmen deutlich mehr Blätter als weibliche Palmen produzieren. Die Ernte von Blättern hat nur geringe Auswirkung auf die Produktion von Blättern, reduziert die Fruchtproduktion jedoch erheblich. Nachhaltige Nutzung und Vermarktung der beiden teilweise exklusiven und lokal konkurrierenden Produkte *tagua* (pflanzliches Elfenbein) und cade (Blätter für Dächer) müssen sorgfältig durchdacht werden. Einsatz nicht nachhaltiger Praktiken in der Ernte von Samen und Blättern, mäßig effiziente Regeneration von Populationen auf Weideland und unzureichende Verfügbarkeit hoch-qualitativer Rohstoffe für die verarbeitende Industrie repräsentieren die größten Herausforderungen bei der nachhaltigen Nutzung dieser wertvollen Palmart in der Zukunft.

Kapitel 5 & 6 befassen sich mit botanischen und ökologischen Aspekten und Hintergründen zur Nachhaltigkeit in der Bewirtschaftung wilder Bestände einer Medizinalpflanze aus den peruanischen Anden (*Krameria lappacea*) und den Auswirkungen der Wildernte auf den Zustand einzelner Populationen.

In Kapitel 5 werden die Ergebnisse zur Studie Parasitismus und Haustoriumanatomie von *Krameria lappacea* (*Ratanhia*, Krameriaceae) präsentiert. *Ratanhia* ist eine gefährdete, übersammelte Heilpflanze aus den Halbwüsten in den südamerikanischen Anden und wird destruktiv aus der Natur geerntet. Die vorliegende Studie untersucht die Anwesenheit oder Abwesenheit von Hemiparasitismus, das Wirtspflanzenspektrum sowie Morphologie und Anatomie der Haustorien. Hemiparasitismus konnte bestätigt werden, 106 haustoriale Verbindungen zu 18 Wirtspflanzenarten aus 17 Gattungen und 12 Pflanzenfamilien wurden dokumentiert. Durch die Ergebnisse dieser Studie wurde die Anzahl von bekannten Wirtspflanzen für *Krameria* verdoppelt und um vier zusätzliche Angiospermefamilien erweitert. *K. lappacea* ist insgesamt ein sehr generalistischer Hemiparasit und hat Einfluss auf die Leistungsfähigkeit der meisten mehrjährigen Arten in ihrem Lebensraum. Übersammlung betrifft daher wahrscheinlich die gesamte Vegetation. Aufgrund der speziellen ökologischen und wirtschaftlichen Bedeutung dieser Art sollten Strategien zur Erhaltung und zur nachhaltigen Ernte hohe Priorität haben.

Kapitel 6 befaßt sich mit den Auswirkungen von Wildernte auf den Zustand natürlicher *Krameria lappacea* Populationen und der Frage wie man auf Basis eines besseren biologischen Verständnisses der Art die Wildernte nachhaltiger gestalten kann. *Krameria* Wurzelextrakte werden als vielversprechende Zutaten für diverse kosmetische und pharmazeutische Präparate eingesetzt. Eine kommerzielle Kultivierung dieser Heilpflanze ist jedoch nicht ökonomisch. Die kommerzielle Beschaffung erfolgt hauptsächlich in Peru. Durch destruktive Erntetechniken und umfangreiche Vermarktung werden wilde Bestände dieser Art immer seltener und in einigen Gegenden ist *Ratanhia* bereits erheblich dezimiert (commercially extinct). Die Studie basiert auf den Abundanzdaten aus der Untersuchung von 238 plots (100 qm) in 6 Regionen in Peru und der Zustand der unterschiedlich beeinflußten Populationen wird verglichen und diskutiert. Experimentelle Ergebnisse zur Keimung von *Ratanhia*, die Relevanz für verfügbare Management-Optionen haben, werden vorgestellt. Strategien zur Erhaltung von *K. lappacea* sollten hohe Priorität haben und es wird dringend empfohlen die Umsetzung verfügbarer nachhaltiger Erntepraktiken zu fördern.

In Kapitel 7 werden für beide Themenkomplexe der Arbeit die Kernaussagen der einzelnen Kapitel abschließend zusammengefasst und diskutiert.

Contribution to Chapters

Chapter 2: Brokamp G., Mittelbach M., Valderrama N., Weigend, M. 2010. Gathering data on production and commercialization of palm products. *Ecología en Bolivia* 45(3): 69–84. Author´s contribution: Designed work (together with M. Weigend); prepared forms, manual and data capture table; translated forms, manual and data capture table into Spanish (together with N. Valderrama), tested draft versions in the field in Peru (together with M. Mittelbach); and wrote the manuscript (together with M.Weigend).

Chapter 3: Brokamp G., Valderrama N., Mittelbach M., Grandez-R. C.A., Barfod A.S., Weigend, M. 2011. Trade in Palm Products in Northwestern South America. *The Botanical Review* 77(4): 571–606. Author´s contribution: Designed work (together with M. Weigend), reviewed literature (together with M. Mittelbach and N. Valderrama) and official statistics, prepared figures and wrote the manuscript (together with A.S. Barfod and M. Weigend).

Chapter 4: Brokamp G., Borgtoft Pedersen H., Montúfar R., Jacome J., Weigend M., Balslev H. in prep. Productivity and management of *Phytelephas aequatorialis* (Arecaceae) in Ecuador. To be submitted to *Annals of Applied Biology*. Author´s contribution: Collected field data (in part), performed statistical analyses (in part), prepared figures, and wrote the manuscript (in part).

Chapter 5: Brokamp G., Dostert N., Cáceres-H. F., Weigend M. 2012. Parasitism and haustorium anatomy of *Krameria lappacea* (Dombey) Burdet & B.B. Simpson (Krameriaceae), an endangered medicinal plant. *Journal of Arid Environments* 83: 94–100. Author´s contribution: Designed work (together with M. Weigend), collected field data and material (in part), analyzed material, prepared figures, and wrote the manuscript (together with M.Weigend).

Chapter 6: Brokamp G., Schwarzer C., Dostert N., Cáceres-H. F., Weigend M. in prep. Now, where did all the Rhatanies go? Abundance, seed ecology, and regeneration of *Krameria lappacea* from the Peruvian Andes. To be submitted to *Journal of Arid Environments*. Author´s contribution: Designed work (together with M. Weigend), designed glasshouse experiments (together with M. Weigend), collected data in the field (in part) and glasshouse, performed statistical analyses, prepared figures, and wrote the manuscript (together with M. Weigend).

Publication list

Weigend M., Brokamp G., Kirbach A., Förther H. 2003. Notas sobre *Heliotropium krauseanum* Fedde, la única especie de Heliotropium sect. Cochranea del Perú, *Arnaldoa* 10(1): 61—74.

Dostert N., Roque J., Brokamp G., Cano A., La Torre M.I., Weigend M. 2009. Factsheet: Botanical Data: *Maca Lepidium meyenii Walp.* GTZ. Proyecto: Desarrollo de monografías botánicas (factsheets) para cinco cultivos peruanos (Documento D38/08-12). Lima, Mayo. 9 pp.

Dostert N., Roque J., Brokamp G., Cano A., La Torre M.I., Weigend M. 2009. Factsheet: Botanical data: *Camu Camu Myrciaria dubia (Kunth) McVaugh.* GTZ. Proyecto: Desarrollo de monografías botánicas (factsheets) para cinco cultivos peruanos (Documento D38/08-6). Lima, Mayo. 9 pp.

Dostert N., Roque J., Brokamp G., Cano A., La Torre M.I., Weigend M. 2009. Factsheet: Botanical Data: *Sacha Inchi Plukenetia volubilis L.* GTZ. Proyecto: Desarrollo de monografías botánicas (factsheets) para cinco cultivos peruanos (Documento D38/08-13). Lima, Mayo. 10 pp.

Dostert N., Roque J., Brokamp G., Cano A., La Torre M.I., Weigend M. 2009. Factsheet: Botanical Data: *Tara Caesalpinia spinosa (Molina) Kuntze.* GTZ. Proyecto: Desarrollo de monografías botánicas (factsheets) para cinco cultivos peruanos (Documento D38/08-11). Lima, Mayo. 9 pp.

Dostert N., Roque J., Brokamp G., Cano A., La Torre M.I., Weigend M. 2009. Factsheet: Botanical Data: *Yacon Smallanthus sonchifolius (Poepp.) H. Rob.* GTZ. Proyecto: Desarrollo de monografías botánicas (factsheets) para cinco cultivos peruanos (Documento D38/08-15). Lima, Mayo. 14 pp.

Brokamp G., Mittelbach M., Valderrama N., Weigend M. 2010. Gathering data on production and commercialization of palm products. Obtención de datos sobre producción y comercialización de productos de palmas. *Ecología en Bolivia* 45(3): 69-84, Diciembre 2010. ISSN 1605-2528.

Brokamp G., Valderrama N., Mittelbach M., Grandez-R. C.A., Barfod A.S., Weigend M. 2011. Trade in Palm Products in Northwestern South America. *The Botanical Review* 77(4): 571-606.

Brokamp G., Valderrama N., Weigend M. 2011. *Palm conservation through use and commercialization?* In: The World of Palms, Lack H.W., Baker W.J. (eds.) – exhibition catalogue, pp. 42-43. BGBM press, Botanischer Garten und Botanisches Museum Berlin-Dahlem (BGBM), FU Berlin, Germany.

Luebert F., Brokamp G., Wen J., Weigend M., Hilger H.H. 2011. Phylogenetic relationships and morphological diversity in Neotropical *Heliotropium* (Heliotropiaceae). *Taxon* 60(3): 663-680.

Montúfar R.J., Brokamp G. 2011. Palmeras aceiteras del Ecuador: estado del arte en la investigación de nuevos recursos oleaginosos provenientes del bosque tropical. *Revista Ecuatoriana de Medicina y Ciencias Biológicas* 32: 93-118.

Brokamp G. 2011. Book review: Galeano, G. & Bernal, R. (2010) Palmas de Colombia: Guía de campo. ICN-UN, Bogotá, Colombia. 688 pp. *ECOTROPICA* 17: 104–105.

Brokamp G., Dostert N., Cáceres-H. F., Weigend M. 2012. Parasitism and haustorium anatomy of *Krameria lappacea* (Dombey) Burdet & B.B. Simpson (Krameriaceae), an endangered medicinal plant. *Journal of Arid Environments* 83: 94–100.

Weigend M., Luebert F., Selvi F., Brokamp G., Hartmut H.H. 2013. Multiple origins for Hound's tongues (*Cynoglossum* L.) and Navel seeds (*Omphalodes* Mill.) – The phylogeny of the borage family (Boraginaceae s.str.). *Molecular Phylogenetics and Evolution* 68(3): 604–618.

Congress contributions

Talks

Weigend M., Brokamp G. 2009. *Sustainability in the real world – wild harvest of palm products in the Neotropics.* EUNOPS 2009, Royal Botanic Gardens Kew, UK.

Brokamp G., Mittelbach M., Valderrama N., Weigend M. 2011. *Trade in Palm products in Northwestern South America – bad future prospects for sustainable resource extraction.* BioSystematics 2011, Berlin, Germany.

Brokamp G., Mittelbach M., Valderrama N., Weigend M. 2011. *Value chains and sustainability in the trade with palm products in NW South America.* EUNOPS 2011, Komarov Botanical Institute, St. Petersburg, Russia.

Brokamp G., Mittelbach M., Valderrama N., Weigend M. 2011. *Obtención y evaluación de datos sobre la comercialización de los productos de palmeras.* FP7-PALMS Symposium, Leticia, Colombia.

Weigend M., Mittelbach M., Brokamp G. 2011. *Neotropical Palm Products – Underutilised ressources for marginal lands.* Tropentag 2011 "Development on the margin", Bonn, Germany.

Weigend M., Mittelbach M., Brokamp G. 2012. *Palmito und Aguaje - Tausend Nutzen für neuweltliche Palmen.* 12.01.2012 in the series „Regenwald - Schatzkammer des Lebens" at the Zoologisches Forschungsmuseum Alexander König, Bonn, Germany.

Brokamp G., Isaza C., Pintaud J.-C., Montúfar R., Dussert S., Weigend M. 2012. *Detecting adulteration and population differentiation by PCA and Cluster analysis with fatty acid composition data of Oenocarpus bataua (Arecaceae) mesocarp oil from NW South America.* 13. Jahrestagung der Gesellschaft für Biologische Systematik, Berlin, Germany.

Brokamp G., Baldassari D., Cevallos D., Weigend M., Balslev H. 2012. *Brooms made from fibres of Aphandra natalia – insights into a typical palm product from NW South America.* EUNOPS 2012, Centro Studi e Ricerche per le Palme Association, Sanremo, Italy.

Brokamp G., Montúfar R., Jacome J., Baldassari D., Weigend M. 2013. *Causes of resource limitation in the trade with tagua (Phytelephas aequatorialis).* EUNOPS 2013, Aarhus University, Aarhus, Denmark.

Poster

Brokamp G., Weigend M., Dostert N. 2005. *Ecology and conservation of hemiparasitic Krameria lappacea (Dombey) Burdet & B.B. Simpson.* 18. Jahrestreffen der Gesellschaft für Tropenökologie, Berlin, Germany.

Luebert F., Brokamp G., Weigend M., Hilger H.H. 2009. *Phylogeny and age estimates of the South American Heliotropium (Heliotropiaceae).* 60° Congresso Nacional de Botânica, Feria de Santana, Brasil.

Brokamp G., Mittelbach M., Weigend M. 2010. *Trade, value chain and legal aspects for two palm products – a case study near Iquitos (Peru).* Palms 2010, International Symposium on the Biology of the Palm Family, Montpellier, France.

Brokamp G., Montúfar R.J., Pintaud J.-C., Dussert S., Weigend M. 2011. *Native palm oils from NW South America – FA composition of Ecuadorean palm oils and a case study on commercially available mesocarp oil from Oenocarpus bataua (Arecaceae).* Botanikertagung 2011 – Diversity makes the difference, Berlin, Germany.

References

Acosta-Solis M. 1944. *La Tagua.* Publicaciones MAS, Quito.

Acosta-Solis M. 1948. Tagua or vegetable ivory – A forest product of Ecuador. *Economic Botany* 2: 46–57.

Adams R. 1999. Germination of Callitris seeds in relation to temperature, water stress, priming, and hydration–dehydration cycles. *Journal of Arid Environments* 43: 437–448.

Adwan G., Salameh Y., Adwan K., Barakat A. 2012. Assessment of antifungal activity of herbal and conventional toothpastes against clinical isolates of Candida albicans. *Asian Pacific Journal of Tropical Biomedicine* 2: 375–379.

Aguirre Z.M., Madsen J.E., Cotton E., Balslev H. (eds.). 2002. *Botánica austroecuatoriana.* Estudio de los recursos vegetales de El Oro, Loja y Zamora Chinchipe. Ediciones Abya-Yala,Quito, Ecuador.

Albán J., Millán B., Kahn F. 2008. Situación actual de la investigación etnobotánica sobre palmeras de Perú. *Revista Peruana De Biología* 15(1): 133–142.

Altieri M.A., Anderson M.K., Merrick L.C. 1987. Peasant Agriculture and the Conservation of Crop and Wild Plant Resources.*Conservation Biology*1(1): 49–58.

Anderson P.J. 1998. *Using ecological and economic information to determine sustainable harvest levels of a plant population.* Pp. 137–155. In: Wollenberg E., Ingles A. (eds.), Incomes from the Forest. Methods for the Development and Conservation of Forest Products for Local Communities. CIFOR – Center for International Forestry Research. IUCN.

Anderson P.J. 2004. The social context for harvesting *Iriartea deltoidea* (Arecaceae). *Economic Botany* 58: 410–419.

Anderson P.J., Putz F.E. 2002. Harvesting and conservation: Are both possible for the palm, *Iriartea deltoidea? Forest Ecology and Management* 170: 271–283.

Angiosperm Phylogeny Group. 2009. An update of the angiosperm phylogeny group classification for the orders and families of flowering plants: APG III. *The Botanical Journal of the Linnean Society* 161: 105–121.

Anonymous. 1999. *Estadisticas de Exportación y Ventas Internas de Productos Forestales a Nivel Nacional.* Gestión 1998. Cámara Nacional Forestal (CNF), Santa Cruz.

Anonymous. 2000. *Non-wood forest products study for Mexico, Cuba and South America.* Forest resources assessment programme.Working paper 11. Forestry Department, Food and Agriculture Organization of the United Nations. FAO, Rome. October 1999.

Anonymous. 2002. Rainforest Conservation Fund. Species Data Sheets, Agroforestry & Ethnobotany. *Phytelephas* spp. (*Tagua*). (available online: www.rainforestconservation.org/agroforestry-ethnobotany/agroforestry-ethnobotany/phytelephas-spp-tagua – accessed 10.11.2010).

Anonymous. 2005. *Plan de Manejo Forestal de Mauritia flexuosa "aguaje"* 2005–2009. Reserva Nacional Pacaya Samiria. Comité de Manejo de Palmeras (COMAPA) "Veinte de Enero". Elaborado con apoyo técnico de ProNaturaleza. Iquitos, Peru. pp 52.

Anonymous. 2009. Centro de Información e Inteligencia Comercial (CICO). Perfiles de Producto. *Perfil del Palmito.* Ecuador.

Anonymous. 2010a. *Perfil de Mercado Palmito.* Instituto Boliviano de Comercio Exterior. Bolivia (IBCE). pp 23.

Anonymous. 2010b. *Panel on Dietetic Products, Nutrition and Allergies (NDA):* Scientific opinion on the substantiation of health claims related to various food(s)/food constituent(s) and protection of cells from premature aging, antioxidant activity, antioxidant content and antioxidant properties, and protection of DNA, proteins and lipids from oxidative damage pursuant to article 13(1) of regulation (EC) No 1924/20061. European Food Safety Authority (EFSA), Parma, Italy. EFSA Journal 8(2): 63. Art. 1489.

Anonymous. 2012. *Informe situacional de la cadena de palma.* Ministerio de Agricultura, Ganadería, Acuacultura y Pesca. Subsecretaría de Comercialización, Ecuador.

Anthelme F., Lincango J., Gully C., Duarte N., Montúfar R. 2011. How anthropogenic disturbances affect the resilience of a keystone palm tree in the threatened Andean cloud forest? *Biological Conservation,* 144(3): 1059–1067.

Anthony R.G. 1976. Influence of drought on diets and numbers of desert deer. Journal of *Wildlife Management* 40: 140–144.

Anthony R.G., Smith N.S. 1977. Ecological relationships between mule deer and white-tailed deer in southeastern Arizona. *Ecological Monographs* 47: 255–277.

Artini M., Papa R., Barbato G., Scoarughi G.L., Cellini A., Morazzoni P., Bombardelli E., Selan L. 2012. Bacterial biofilm formation inhibitory activity revealed for plant derived natural compounds. *Bioorganic & Medicinal Chemistry* 20, 920–926.

Angiosperm Phylogeny Group, 2009. An update of the angiosperm phylogeny group classification for the orders and families of flowering plants: APG III. *The Botanical Journal of the Linnean Society* 161: 105–121.

Asmussen C.B., Dransfield J., Deickmann V., Barfod A.S., Pintaud J.-C., Baker W.J. 2006. A new subfamily classification of the palm family (Arecaceae): evidence from plastid DNA phylogeny. *Botanical Journal of the Linnean Society* 151(1): 15–38.

Backes P., Irgang B. 2004. *Mata Atlântica: As Árvores e a Paisagem*, Porto Alegre: Paisagem do Sul. pp 393.

Balick M.J. 1985. *Current Status of Amazonian Oil Palms.*Reprinted from: Oil Palms and other Oilseeds of the Amazon. Pesce C., Johnson, D.V. (eds.). Reference publications, Inc.

Balick M.J. 1986. Systematics and economic botany of the *Oenocarpus–Jessenia* (Palmae) complex. *Advances in Economic Botany* 3: 1–140.

Balick M.J. 1992. *Jessenia* y *Oenocarpus*: Palmas Aceiteras Neotropicales Dignas de ser Domesticadas. FAO, Estudio para la Producción y Protección Vegetal 88. Roma. pp 180.

Balick M.J., Beck H.T. (eds.). 1990. *Useful palms of the world*. A synoptic bibliography. Columbia University Press, New York.

Balick M.J., Gershoff S.N. 1981. Nutritional evaluation of the *Jessenia bataua* palm: Source of high quality protein and oil from tropical America. *Economic Botany* 35: 261–271.

Balslev H. 2011. Palm harvest impacts in north-western South America. *The Botanical Review* 77(4): 370–380.

Balslev H., Barfod A. 1987. Ecuadorian palms—an overview. *Opera Botanica* 92: 17–35.

Balslev H., Grandez C., Paniagua Zambrana N.Y., Møller A.L. Hansen S.L.. 2008. Palmas (Arecaceae) útiles en los alrededores de Iquitos, Amazonia Peruana. *Revista Peruana de Biología* 15(1): 121–132.

Balslev H., Eiserhardt W., Kristiansen T., Pedersen D., Grandez C. 2010a. Palms and palm communities in the upper Ucayali river valley—a little-known region in the Amazon basin. *Palms* 54: 57–72.

Balslev H., Navarrete H., Paniagua-Zambrana N., Pedersen D., Eiserhardt W., Kristiansen T. 2010b. El uso de transectos para el estudio de comunidades de palmas. *Ecología en Bolivia* 45(3): 8–22.

Barfod A.S. 1989. The rise and fall of vegetable ivory. *Principes* 33: 181–190.

Barfod A.S. 1991a. A monographic study of the subfamily Phytelephantoideae (Arecaceae). *Opera Botanica* 105: 1–73.

Barfod A.S. 1991b. *Usos pasados, presente y futuros de las palmas Phytelephantoidees.* In Las Plantas y El Hombre, pp. 23–46. Rios M., Borgtoft Pedersen, H. (eds.), Quito, Ecuador: Abya-Yala and Herbario QCA.

Barfod A.S., Balslev H. 1988. The use of palms by the Cayapas and Coaiqueres on the coastal plain of Ecuador. *Principes* 32: 29–42.

Barfod, A. S. & L. P. Kvist. 1996. Comparative ethnobotanical studies of the Amerindian groups in coastal Ecuador. *BiologiskeSkrifter* 46: 5–166.

Barfod, A. S., B. Bergmann & H. Borgtoft Pedersen. 1990. The vegetable ivory industry: Surviving and doing well in Ecuador. *Economic Botany* 44: 293–300.

Barrera-Z. V.A., Torres-R. M.C., Ramírez-P. D.S. 2008. *Habilitación, Uso y Manejo Sostenible de Materias Primas Vegetales y Ecosistemas Relacionados con la Producción Artesanal en Colombia.* Protocolo para Producción Sostenible de Artesanías en Palma Estera (*Astrocaryum malybo*) en el Cesar. Ministerio de Comercio, Industria y Turismo. Artesanias de Colombia S.A. Bogotá. pp 57.

Bastian O. 2013. The role of biodiversity in supporting ecosystem services in Natura 2000 sites. *Ecological Indicators* 24, 12–22.

Bates D.M. 1988. Utilization pools: A framework for comparing and evaluating the economic importance of palms. *Advances in Economic Botany* 6: 56–64.

Baumgartner L., Schwaiger S., Stuppner H. 2011. Quantitative analysis of anti-inflammatory lignan derivatives in Ratanhiae radix and its tincture by HPLC–PDA and HPLC–MS. *Journal of Pharmaceutical and Biomedical Analysis* 56: 546–552.

BCE. 2012. Banco Central del Ecuador. *Database of export statistics* (Comercio exterior). URL http://www.portal.bce.fin.ec/vto_bueno/seguridad/ComercioExteriorEst.jsp (accessed on 20.06.2012).

Beck M.J., Vander Wall S.B. 2010. Seed dispersal by scatter-hoarding rodents in arid environments. *Journal of Ecology* 98: 1300–1309.

Belcher B. 2003. What isn't a NTFP? *International Forestry Review* 5(2):161–167.

Belcher B., Ruiz Perez M., Achdiawan R. 2003. *Global Patterns and Trends in NTFP Development.* Paper presented to the international conference "Rural Livelihoods, Forests, and Biodiversity", Bonn, Germany, May 19-23, 2003.

Belcher B., Schreckenberg K. 2007. Commercialisation of Non-timber Forest Products: A Reality Check. *Development Policy Review* 25(3): 355–377.

Bennett B.C., Alarcón R., Cerón C. 1992. The ethnobotany of *Carludovica palmata* Ruíz&Pavón (Cyclanthaceae) in Amazonian Ecuador. *Economic Botany* 46: 233–240.

Bennett B.C. 2002. Forest products and traditional peoples: Economic, biological, and cultural considerations. *Natural Resources Forum* 26: 293–301.

Bereau D., Benjelloun-Mlayah B., Delmas M. 2001. *Maximiliana maripa* Drude mesocarp and kernel oils: Fatty acid and total tocopherol compositions. *Journal of the American Oil Chemists' Society* 78: 213–214.

Bereau D., Benjelloun-Mlayah B., Banoub J., Bravo R. 2003. FA and unsaponifiable composition of five Amazonian palm kernel oils. *Journal of the American Oil Chemists' Society* 80: 49–53.

Bernal R. 1992. *Colombian palm products.* Pp 158–172. In: M. Plotkin, Famolare L. (eds.). Sustainable harvest and marketing of rainforest products. Island Press, Washington.

Bernal R., Torres C., García N., Isaza C., Navarro J., Vallejo M.I., Galeano G., Balslev H. 2011. Palm Management in South America. *The Botanical Review* 77(4): 607–646.

Bernard S.R., Brown K.F. 1977. Distribution of Mammals, Reptiles, and Amphibians by BLM Physiographic Regions and A.W. Kuchler's Associations for The Eleven Western States.Denver, CO. U.S..Department of the Interior, Bureau of Land Management. *Technical Note* 301, 1-169.

Berndes G., Hoogwijk M., van den Broek R. 2003. The contribution of biomass in the future global energy supply: a review of 17 studies. *Biomass and Bioenergy* 25: 1–28.

Bianchi C. 1982. *Artesanías y Técnicas Shuar.* Ediciones Mundo Shuar, Quito.

Blanco-Metzler A., Campos M.M., Piedra M.F., Mora U.J., Montero C.M., Fernandez P.M. 1992. Pejibaye palm fruit contribution to human nutrition. *Principes* 36: 66–69.

Bochet E., García-Fayos P., Alborch B., Tormo J. 2007. Soil water availability effects on seed germination account for species segregation in semiarid roadslopes. *Plant Soil* 295: 179–191.

Bodmer R.E., Penn J.W., Puertas P.E., Moya L.I., Fang T.G. 1997. *Linking conservation and local people through sustainable use of natural resources: Community-based management in the Peruvian Amazon.* Pp 315–358. In: Freese C.H. (ed.). Harvesting wild species: Implications for biodiversity conservation. John Hopkins University Press, Baltimore.

Bora P.S., Narain N., Rocha R.V.M., De Oliveira M.A.C., De Azevedo R. 2001. Characterización de las fracciones proteicas y lipídicas de la pulpa y semillas de Tucuma (*Astrocaryum vulgare* Mart.). *Ciência e Tecnologia de Alimentos* 3: 111–116.

Borchsenius F., Borgtoft Pedersen H., Balslev H. 1998. *Manual to the Palms of Ecuador.* AAU Reports 37, Aarhus, Denmark.

Borchsenius F., Moraes-R M.R. 2006. *Diversidad y usos de palmeras andinas* (Arecaceae). Pp 412–433. In: Moraes-R M., Ollgaard B., Kvist L.P., Borchsenius F., Balslev H. (eds.). Botánica Económica de los Andes Centrales. Universidad Mayor de San Andrés, La Paz.

Borgtoft Pedersen H. 1992. Uses and management of *Aphandra natalia* (Palmae) in Ecuador. *Bulletin de l'Institut Français d'Etudes Andines* 21: 741–753.

Borgtoft Pedersen H. 1994. Mocora palm-fibers: Use and management of *Astrocaryum standleyanum* (Arecaceae) in Ecuador. *Economic Botany* 48: 310–325.

Borgtoft Pedersen H. 1995. Predation of *Phytelephas aequatorialis* seeds („vegetable ivory") by the bruchid beetle Caryoborus chiriquensis. *Principes* 39(2): 89–94.

Borgtoft Pedersen H. 1996. Production and harvest of fibers from *Aphandra natalia* (Palmae) in Ecuador. *Forest Ecology and Management* 80: 155–161.

Borgtoft Pedersen H., Balslev H. 1990. *Ecuadorean palms for agroforestry.* AAU Reports 23: 1–122.

Borgtoft Pedersen H., Balslev H. 1992. *The economic botany of Ecuadorean palms*. Pp 173–191. In: Plotkin M., Famolare L. (eds.). Sustainable harvest and marketing of rain forest products. Island Press, Washington.

Borgtoft Pedersen H., Balslev H. 1993. *Palmas Utiles, Especies Ecuatorianas para Agroforesteria y Extractivismo*. Abya-Yala, Quito, Ecuador.

Borgtoft Pedersen H., Skov F. 2001. Mapping palm extractivism in Ecuador using pair-wise comparisons and bioclimatic modeling. *Economic Botany* 55: 63–71.

Bovi M.L.A., de Castro A. 1993. *Assai*. FAO Working paper. In: Clay J.W., Clement C.R (eds.), Selected Species and Strategies to Enhance Income Generation from Amazonian Forests. Food and Agriculture Organization of the United Nations, Rome, May 1993. (available online: http://www.fao.org/docrep/V0784E/v0784e00.HTM—accessed 20.10.2011)

Boyer K. 1992. *Palms and Cycads Beyond the tropics*. Palm and Cycad Societies of Australia (PACSOA).150 pp.

Bresinsky A., Körner C., Kadereit J.W., Neuhaus G., Sonnewald U. 2008. *Strasburger - Lehrbuch der Botanik*. Begründet von E. Strasburger, 36. Aufl. Spektrum Akademischer Verlag, Heidelberg, ISBN 978-3827414557.

Brokamp G., Mittelbach M., Weigend M. 2010a. Trade, value chain and legal aspects for two palm products – a case study near Iquitos (Peru). Poster: Palms 2010, International Symposium on the Biology of the Palm Family. Montpellier, France. (available online: http://fp7-palms.org/component/content/article/18/188-posters-presented-by-fp7-palms-participants.html—accessed 20.11.2010)

Brokamp G., Mittelbach M., Valderrama N., Weigend M. 2010b. Obtención de datos sobre producción y comercialización de productos de palmas. *Ecología en Bolivia* 45(3): 69–84.

Brokamp G., Valderrama N., Mittelbach M., Grandez C., Barfod A.S., Weigend M. 2011. Trade in Palm Products in Northwestern South America. *The Botanical Review* 77(4): 571–606.

Brokamp G., Dostert N., Cáceres Huamaní F., Weigend M. 2012. Parasitism and haustorium anatomy of *Krameria lappacea* (Dombey) Burdet & B.B.Simpson (Krameriaceae), an endangered medicinal plant from the Andean deserts. *Journal of Arid Environments* 83: 94–100.

Brondízio E.S. 2008. The Amazonian caboclo and the açai palm—forest farmers in the global market. *Advances in Economic Botany* 16: 169–241.

Brown A.L. 1950. Shrub invasion of southern Arizona desert grassland. *Journal of Range Management* 3: 172–177.

Bussmann R.W., Sharon D. 2006. Traditional medicinal plant use in Northern Peru: tracking two thousand years of healing culture. *Journal of Ethnobiology and Ethnomedicine* 2: 47.

Bussmann R.W., Glenn A., Sharon D. 2010. Antibacterial activity of medicinal plants of Northern Peru – can traditional applications provide leads for modern science? *Indian Journal of Traditional Knowledge* 9: 742–753.

Bussmann R.W., Glenn A., Sharon D., Chait G., Díaz D., Pourmand K., Jonat B., Somogy S., Guardado G., Aguirre C., Chan R., Meyer K., Rothrock A., Townesmith A. 2011. Proving that traditional knowledge works: The antibacterial activity of Northern Peruvian medicinal plants. *Ethnobotany Research & Applications* 9: 67–96.

Callaway R.M., Pennings S.C. 1998. Impact of a parasitic plant on the zonation of two salt marsh perennials. *Oecologia* 114: 100–105.

Cannon W.A. 1910. *The root habits and parasitism of Krameria canescens Gray*. Pp. 5–24 In: Macdougal D.T., Cannon W.A. (eds.), The Conditions of Parasitism in Plants, vol. 129. Publications of the Carnegie Institute of Washington.

Cannon W.A. 1911. *The Root Habits of Desert Plants*, vol. 131. Publications of the Carnegie Institute of Washington 67–69.

Carini M., Aldini G., Orioli M., Facino R.M. 2002. Antioxidant and photoprotective activity of a lipophilic extract containing neolignans from *Krameria triandra* roots. *Planta Medica* 68: 193–197.

Castaño N., Cárdenas D., Otavo E. 2007. *Ecología, Aprovechamiento y Manejo Sostenible de Nueve Especies de Plantas del Departamento del Amazonas, Generadoras de Productos Maderables y No Maderables*. Instituto Amazónico de Investigaciones Científicas – SINCHI. Corporación para el Desarrollo Sostenible del sur de la Amazonia, CORPOAMAZONIA, Bogotá.

Chandrasekharan C. 1995. *Terminology, definition and classification of forest products other than wood*. Pp. 345–380 In: FAO. Report of the International expert consultation on non-wood forest products, Yogyakarta,Indonesia. 17–27 January 1995. Non-wood forest products no. 3, FAO, Rome.

Clement C.R., Arkcoll D.B. 1985. *El Bactris gasipaes como cultivo oleaginoso potencial y necesidades de domesticación.* Pp. 16–20 In: Forero L.E. (ed.). Semmano-Taller sobre Oleaginosas Promisorias. Pirb/Interciencia, Bogotá.

Clement C.R., Mora-Urpí J.E. 1987. Pejibaye palm (*Bactris gasipaes*, Arecaceae): Multi-use potential for the lowland humid tropics. *Economic Botany* 41: 302–311.

Clement C.R., Weber J.C., van Leeuwen J., Astorga-D C., Cole D.M., Arévalo-L L.A., Argüello H. 2004. Why extensive research and development did not promote use of peach palm fruit in Latin America. *Agroforestry Systems* 61: 195–206.

Cohen A.L., Shaykh M. 1973. Fixation and dehydration in the preservation of surface structure in critical point drying of plant material. *Scanning Electron Microscopy* 6: 371–378.

Coïsson D., Travaglia F., Piana G., Capasso M., Arlorio M. 2005. *Euterpe oleracea* juice as a functional pigment for yogurt. *Food Research International* 38: 893–897.

Coomes O.T. 1996. Income formation among Amazonian peasant households in northeastern Peru: Empirical observations and implications for market-oriented conservation. *Yearbook, Conference of Latin Americanist Geographers* 22: 51–64.

Coomes O.T. 2004. Rain forest 'conservation-through-use'? Chambira palm fibre extraction and handicraft production in a land-constrained community, Peruvian Amazon. *Biodiversity and Conservation* 13: 351–360.

Coomes O.T., Barham B.L. 1997. Rain forest extraction and conservation in Amazonia. *The Geographical Journal* 163: 180–188.

Couvreur T.L.P., Hahn W.J., De Granville J.-J., Pham J.-L., Ludeña B., Pintaud J.-C. 2007. Phylogenetic relationships of the cultivated Neotropical palm Bactrisgasipaes (Arecaceae) with its wild relatives inferred from chloroplast and nuclear DNA polymorphisms. *Systematic Botany* 32: 519–530.

Daems W.F. 1981. Radix Ratanhiae – die Droge mit einer gesicherten Geschichte. *Deutsche Apotheker Zeitung* 121: 46–52.

Daily G.C. 1997. *Introduction: What are Ecosystem Services?* In: Daily G.C. (ed.) Nature's Services: Societal Dependence on Natural Ecosystems. Island Press. Washington, DC.

De Beer J.H., McDermott M.J.1989. *The economic value of non-timber forest products in Southeast Asia with emphasis on Indonesia, Malaysia and Thailand.* 175 pp.

De la Cruz S.H., Zevallos P.,Vilcapoma G.S. 2005. Status de conservacíon de las especies vegetales silvestres de uso tradicional en la provincia de Canta, Lima-Perú. *Ecología Aplicada* 4: 9–16.

De la Cruz H., Vilcapoma G., Zevallos P.A. 2007. Ethnobotanical study of medicinal plants used by the Andean people of Canta, Lima, Peru. *Journal of Ethnopharmacology* 111: 284–294.

De la Torre, L., Navarrete H., Muriel-M. P., Macía M.J., Balslev H. (eds.). 2008. *Enciclopedia de las Plantas Útiles del Ecuador.* Herbario QCA de la Escuela de Ciencias Biológicas de la Pontificia Universidad Católica del Ecuador & Herbario AAU del Departamento de Ciencias Biológicas de la Universidad de Aarhus. Quito & Aarhus.

De la Torre L., Valencia R., Altamirano C., Munk Ravnborg H. 2011. Legal and administrative regulation of palms and other NTFPs in Colombia, Ecuador, Peru and Bolivia. *The Botanical Review* 77(4): 327–369.

Del Castillo D., Otárola E., Freitas L. 2006. *Aguaje, la Maravillosa Palmera de la Amazonía.* IIAP, Instituto de Investigaciones de la Amazonía Peruana, Wust W.H. (ed.). Ediciones Wust. 51.

Delgado C., Couturier G., Kember M. 2007. *Mauritia flexuosa* (Arecaceae: Calamoideae), an Amazonian palm with cultivation purposes in Peru. *Fruits* 62: 157–169.

De Oliveira, M. K. S., H. E. Martinez-F, J. S. De Andrade, M. G. Garnica-R & Y. K. Chang. 2006. Use of pejibaye flour (*Bactris gasipaes* Kunth) in the production of food pastas. *International Journal of Food Science and Technology* 41: 933–937.

Dewis J., Freitas F. 1984. *Métodos físicos y químicos de análisis de suelos y aguas.* FAO ONU, Rome, 252 pp.

Dodson C.H., Gentry A.H. 1991. Biological extinction in western Ecuador. *Annals of the Missouri Botanical Garden* 78: 273–295.

Dransfield J., Uhl N.W., Asmussen C.B., Baker W.J., Harley M.M., Lewis C.E. 2008. *Genera Palmarum II – the evolution and classification of palms.* Richmond, UK.

Duarte N., Montúfar R. 2012. Effect of leaf harvest on wax palm (*Ceroxylon echinulatum* Galeano) growth, and implications for sustaiable management in Ecuador. *Tropical Conservation Science* 5: 340–351.

Dubois V., Breton S., Lindera M., Fannia J., Parmentiera M.. 2007. Fatty acid profiles of 80 vegetable oils with regard to their nutritional potential. *European Journal of Lipid Science and Technology* 109: 710–732.

Duivenvoorden J.F., Balslev H., Cavalier J., Grandez C., Tuomisto H., Valencia R. (eds.). 2001. *Evaluación de recursos vegetales no maderables en la Amazonía noroccidental.* Institute for Biodiversity and Ecosystem Management (IBED)-Paleo-Actuo-Ecology, University of Amsterdam, Amsterdam , The Netherlands .

Echlin P. 1978. Coating techniques for scanning electron microscopy and X-ray microanalysis.*Scanning Electron Microscopy* 1: 108–132.

Ehrlich P.R., Holdren J.P. 1971. Impact of Population Growth. *Science, New Series* 171(3977): 1212–1217.

Elliot J.M. 1977. *Some methods for the statistical analysis of samples of benthic invertebrates.* Freshwater Biological Association. Scientific Publication 25: 1–160.

Endress B.A., Horn C.M., Gilmore M.P. 2013. *Mauritia flexuosa* palm swamps: Composition, structure and implications for conservation and management *Forest Ecology and Management* 302; 346–353.

Falconer J. 1990. *The major significance of "minor" forest products: the local use and value of forests in the West African humid forest zone.* Community Forestry Notes. FAO, Rome.

FAO. 2005. Food and Agriculture Organization. *Proceedings: third expert meeting on harmonizing forest-related definitions for use by various stakeholders*, FAO, Rome, 17–19 January 2005.

FAO. 2010. Food and Agriculture Organization. T*he State of Food Insecurity in the World 2010.* http://www.fao.org/docrep/013/i1683e/i1683e.pdf

FAOSTAT. 2011. Food and Agricultural Organization. *Statistical database.* URL http://faostat.fao.org/site/339/default.aspx (acessed 10. 11.2011).

Feldman D. Gagnon J. Hofmann R. Simpson J. 1988. StatView™ SE+ Graphics. Version 1.02. Abacus Concepts Inc., Berkeley, California.

Flores C.F., Ashton P.M.S. 2000. Harvesting impact and economic value of *Geonoma deversa*, Arecaceae, an understory palm used for roof thatching in the Peruvian Amazon. *Economic Botany* 54: 267–277.

Flores J., Briones O. 2001. Plant life-form and germination in a Mexican inter-tropical desert: effects of soil water potential and temperature. *Journal of Arid Environments* 47: 485–497.

Frohne D. 2006. *Heilpflanzenlexikon*. Ein Leitfaden auf wissenschaftlicher Grundlage, 8. neu bearbeitete Auflage. Deutscher Apotheker Verlag, Stuttgart.

Froud-Williams R.J., Chancellor R.J., Drennan D.S.H. 1984. The effects of seed burial and soil disturbance on emergence and survival of arable weeds in relation to minimal cultivation. *Journal of Applied Ecology* 21: 629–641.

Galeano G., Bernal R. 1987. *Palmas del Departamento de Antioquia, Región Occidental*. Universidad Nacional de Colombia, Bogotá.

Galeano G., Bernal R. 2010. *Palmas de Colombia*. Guia de Campo. Editorial Universidad Nacional de Colombia. Instituto de Ciencias Naturales, Facultad de Ciencias – Universidad Nacional de Colombia, Bogotá.

García A., Pinto J.J. 2002. *Diagnóstico de la Demanda del Aguaje en Iquitos*. IIAP. Instituto de Investigaciones de la Amazonía Peruana.

García N., Torres-R. M.C., Valderrama N., Bernal R., Galeano G. 2010. *Astrocaryum fibers for handicraft production in Colombia*. Poster Palms 2010: International Symposium on the Biology of the Palm Family. Montpellier, France. (available online: http://fp7-palms.org/component/content/article/18/188-posters-presented-by-fp7-palms-participants.html—accessed 20.11.2010)

Giannini T.C., Takahasi A., Medeiros M.C.M.P., Saraiva A.M., Alves-dos-Santos I. 2011. Ecological niche modeling and principal component analysis of *Krameria* Loefl. (Krameriaceae). *Journal of Arid Environments* 75: 870–872.

Gilmore M.P., Endress B.A., Horn C. 2013. The socio-cultural importance of *Mauritia flexuosa* palm swamps (*aguajales*) and implications for multi-use management in two Maijuna communities of the Peruvian Amazon. *Journal of Ethnobiology and Ethnomedicine* 9: 29

Gimenes M., Da Lobão C. 2006. A Polinização de *Krameria bahiana* B.B. Simpson (Krameriaceae) por Abelhas (Apidae) na Restinga, BA. *Neotropical Entomology* 35: 440–445.

Goldberg D.E., Turner R.M. 1986. Vegetation change and plant demography in permanent plots in the Sonoran Desert. *Ecology* 67: 695–712.

Gonzales D.A., Perez-Prat Vinuesa E.M., Dkidak A., Asmanidou A. 2012. *United States Patent Application Publication. Pub. No.: US 2012/0071379 A1.* URL http://www.patentstorm.us/applications/20120071379/claims.html (accessed on 20.06.2012).

Govaerts R., Dransfield J. 2005. Pp 223. *World checklist of palms.* Royal Botanic Gardens, Kew.

Govaerts R., Dransfield J., Zona S.F., Hodel D.R., Henderson, A. 2013. *World Checklist of Arecaceae.* Facilitated by the Royal Botanic Gardens, Kew. Published on the Internet; http://apps.kew.org/wcsp/ Retrieved 2013-06-18

Grazi D. 2008. *Ratanhia: von den peruanischen Anden in die Tube.* Schriftliche Arbeit im Rahmen des Zertifikatstudienganges (CAS) Ethnobotanik und Ethnomedizin, Universität Zürich, 21 pp. URL http://www.weiterbildung.uzh.ch/programme/ethnobot/Abschlussarbeiten/RatanhiaDaniaGraziEndversion.pdf (accessed 08.04.2013).

Guel A., Penn J.W. 2009. *Use of the chambira palm (Astrocaryum chambira) in rainforest communities of the Peruvian Amazon.* Pp. 1–26. In: Undergraduate Research and Creative Practice – Student Summer Scholars. Grand Valley State University, Grand Rapids. URL http://scholarworks.gvsu.edu/sss/20 (accessed on 20.11.2010)

Gupta M.P. 2006. *Medicinal plants originating in the Andean high plateau and Central Valles region of Bolivia, Ecuador and Peru.* UNIDO Report. United Nations – Industrial and technology promotion branch: The future of products of the Andean high plateau and Central valleys.

Hanley T.A., Brady W.W. 1977. Feral burro impact on a Sonoran Desert range. *Journal of Range Management* 30: 374–377.

Hanley N., Shogren J., White B. 2007. *Environmental Economics in Theory and Practice,* Palgrave, London.

Hayden P. 1966. Food habits of black-tailed jack rabbits in southern Nevada. *Journal of Mammalogy* 47: 42–46.

Harris J. 2006. *Environmental and Natural Resource Economics: A Contemporary Approach.* Houghton Mifflin Company.

Henderson A. 1995. *The palms of the Amazon.* Oxford University Press, New York. 362 pp.

Henderson A. 2000. *Bactris (Palmae).* Flora Neotropica Monograph 79. New York Botanical Garden, New York.

Henderson A., Galeano G., Bernal R. 1995. *A field guide to the palms of the Americas.* Princeton University Press. Princeton, New Jersey.

Hilgert N.I. 2001. Plants used in home medicine in the Zenta River basin, Northwest Argentina. *Journal of Ethnopharmacology* 76: 11–34.

Hofmann K. 1995. *Schmöllner Knöpfe.* Pp 48. Museum Burg Posterstein. Posterstein, Germany.

Holdridge L.R., Grenke W.C., Hatheway W.H., Liang T., Tosi J.A. 1971. *Forest Environments in Tropical Life Zones: A Pilot Study.* Pergamon, New York.

Holm J. A., Miller C.J., Cropper Jr. W.P. 2008. Population dynamics of the dioecious Amazonian palm *Mauritia flexuosa*: Simulation analysis of sustainable harvesting. *Biotropica* 40: 550–558.

Holm Jensen O., Balslev H. 1995. Ethnobotany of the fiber palm *Astrocaryum chambira* (Arecaceae) in amazonian Ecuador. *Economic Botany* 49: 309–319.

Holmgren C.A., Betancourt J.L., Rylander K.A., Roque J., Tovar O., Zeballos H., Linares E., Quade J. 2001. Holocene vegetation history from fossil rodent middens near Arequipa, Peru. *Quaternary Research* 56: 242–251.

IBM Corp. 2012. IBM SPSS Statistics for Windows, Version 21.0. Armonk, NY.

IPCC. 2001. Intergovernmental Panel on Climate Change. *Climate Change 2001: The Scientific Basis.* http://www.grida.no/climate/ipcc_tar/wg1/figts-22.htm

IPCC. 2007. Intergovernmental Panel on Climate Change. *Climate Change 2007: Synthesis report. Contribution of Working Groups I, II, and III to the Fourth Assessment Report of the Intergovernmental Panel on Climate Change.* IPCC, Geneva,Switzerland.

Isaza-A. C., Bernal R., Howard P. 2010. *Use, production and conservation of palm fibres in South America.* Poster: PALMS 2010, International Symposium on the Biology of the Palm Family. Montpellier, France. (available online http://fp7-palms.org/component/content/article/18/188-posterspresented-by-fp7-palms-participants.html—accessed 20.11.2010)

Isaza C., Bernal R., Howard P. 2013 Use, Production and Conservation of Palm Fiber in South America: A Review. *Journal of Human Ecology* 42(1): 69–93.

Jacobo F.Q., Rojas M.A., Reyes G.I., Pino E.L., Chagman G.P. 2009. Caracterización de aceites, tortas y harinas de frutos de Ungurahui (*Jessenia polycarpa*) y Aguaje (*Mauritia flexuosa* L.) de la Amazonía Peruana. *Revista de la Sociedad Química del Perú* 75: 243–253.

Jahnke H.E. 1982. *Livestock production systems and livestock development in tropical Africa.* Kiel: Wissenschaftsverlag Vauk.

Janer K. J.C. 2002a. *Estudio de Mercado Nacional del Palmito en Colombia.* Prepared for: Chemonics International Inc., Washington, D.C. (Colombia Alternative Development Project). Bogotá, Colombia.

Janer K. J.C. 2002b. *Estudio del Mercado Mundial de Palmito y Subproductos.* Prepared for: Chemonics International Inc., Washington D.C. (Colombia Alternative Development Project). Bogotá, Colombia.

Jatunov S., Quesada S., Díaz C., Murillo E. 2010. Carotenoid composition and antioxidant activity of the raw and boiled fruit mesocarp of six varieties of *Bactris gasipaes*. *Archivos Latinoamericanos de Nutrición* 60: 99–104.

Johnson D.V. & the IUCN/SSC Palm Specialist Group. 1996. *Palms: Their conservation and sustained utilization. Status survey and conservation action plan.* IUCN, Gland. Switzerland and Cambridge, UK.

Johnson D.V. 2011. *Tropical palms.* 2010 revision. Non-Wood Forest Products No. 10/ Rev.1.Food and Agriculture Organization of the United Nations (FAO), Publishing Management Service Information Division, Rome, Italy. 242 pp.

Joshi J., Matthies D., Schmid B. 2000. Root hemiparasites and plant diversity in experimental grassland communities. *Journal of Ecology* 88: 634–644.

Kahn F. 1988. Ecology of economically important palms in Peruvian Amazonia. *Advances in Economic Botany* 6: 42–49.

Kahn F. 1991. Palms as key swamp forest resources in Amazonia. *Forest Ecology and Management* 38: 133–142.

Kahn F., De Granville J.J. 1992. *Palms in forest ecosystems of Amazonia.* Ecological studies 95. Springer, Berlin.

Kahn F., Henderson A. 1999. *An overview of the palms of the várzea in the Amazon region.* Pp 187–193. In: Padoch C., Ayres J.M., Pinedo-Vasquez M., Henderson A. (eds.). Várzea diversity, development, and conservation of Amazonia's white-water flood-plains.The New York Botanical Garden, New York.

Kahn F., Mejía K.M. 1987. Notes on the biology, ecology and use of a small Amazonian palm: *Lepidocaryum tessmannii. Principes* 31: 14–19.

Kaplinsky R., Morris M. 2002. *A Handbook for Value Chain Research.* Report prepared for IDRC.

Kommission E. 1989. *Monographie BGA/BfArM.* Bundesanzeiger, 43., ATC-Code: A01AF.

Koziol M.J., Borgtoft Pedersen H. 1993.) *Phytelephas aequatorialis* in human and animal nutrition. *Economic Botany* 47(4): 401–407.

Kronborg M., Grandez C., Ferreira E., Balslev H. (2008) *Aphandra natalia* (Arecaceae) – a little known source of piassaba fibers from the western Amazon. *Revista Peruana de Biología* 15: 103–113.

Kuch M., Rohland N., Betancourt J.L., Latorre C., Steppan S., Poinar H.N. 2002. Molecular analysis of a 11 700-year-old rodent midden from the Atacama Desert, Chile. *Molecular Ecology* 11: 913–924.

Kuijt J. 1969. *The biology of parasitic flowering plants.* University of California Press, 15 Berkeley, CA.

Kusters K., Achdiawan R., Belcher B., Ruiz Pérez M. 2006. Balancing development and conservation? An assessment of livelihood and environmental outcomes of non timber forest product trade in Asia, Africa, and Latin America. *Ecology and Society* 11(2): 20. URL: http://www.ecologyandsociety.org/vol11/iss2/art20/ (accessed on 10.11.2012)

Kvist L.P., Nebel G. 2001. A review of Peruvian flood plain forests: ecosystems, inhabitants and resource use. *Forest Ecology and Management* 150(1-2): 3–26.

Lange D., Schippmann U. 1997. *Trade Survey of Medicinal Plants in Germany: A Contribution to International Plant Species Conservation.* Bundesamt für Naturschutz, Bonn

Latorre C., Betancourt J.L., Rylander K.A., Quade J. 2002. Vegetation invasions into absolute desert: A 45 000 yr rodent midden record from the Calama–Salar de Atacama basins, northern Chile (lat 22°–24°S). *Geological Society of America Bulletin* 114(3): 349–366.

Lescure J.P., Emperaire L., Franciscon C. 1992. *Leopoldinia piassaba* Wallace (Areacaceae): A few biological and economic data from the Rio Negro region (Brazil). *Forest Ecology and Management* 55: 83–86.

Lévi-Strauss C. 1952. The use of wild plants in tropical South America. *Economic Botany* 6: 252–270.

Linares E., Galeano G., Figueroa Y., García N. 2008. *Fibras Vegetales Empleadas en Artesanías en Colombia.* Universidad Nacional de Colombia, Facultad de Ciencias, Instituto de Ciencias Naturales. Artesanías de Colombia S. A. Bogotá.

Lleras E., Coradin L. 1988. Native neotropical oil palms: State of the art and perspectives for Latin America. *Advances in Economic Botany* 6: 201–203.

Lozada T., de Koning G.H.J., Marché R., Klein A.-M., Tscharntke T. 2007. Tree recovery and seed dispersal by birds: Comparing forest, agroforestry and abandoned agroforestry in coastal Ecuador. *Perspectives in Plant Ecology, Evolution and Systematics* 8: 131–140.

Macía M.J. 2004. Multiplicity in palm uses by the Huaorani of Amazonian Ecuador. *Botanical Journal of the Linnean Society* 144: 149–159.

Macía M.J., Armesilla P.J., Cámara-Leret R., Paniagua-Zambrana N., Villalba S., Balslev H., Pardo de Santayana M. 2011. Palm Uses in Northwestern South America: A Quantitative Review. *The Botanical Review* 77(4): 462–570.

Macuyama-R. W. 2008. Comunidad Campesina de Ribereños Roca Fuerte, *Plan de Manejo Forestal con Fines No Maderables* (Baja Escala – R. J. N° 232-2006-INRENA) de la Comunidad Campesina Roca Fuerte (Cuenca Chambira).

Madriñán S., Schultes R.E. 1995. Colombia's national tree: The wax palm *Ceroxylon quindiuense* and its relatives. *Elaeis* 7: 35–56.

Mantau U., Wong J.L.G., Curl S. 2007. Towards a taxonomy of forest goods and services. *Small-scale Forestry* 6: 391–409.

Manzi M., Coomes O.T. 2009. Managing Amazonian palms for community use: A case of aguaje palm (*Mauritia flexuosa*) in Peru. *Forest Ecology and Management* 257: 510–517.

Marshall E., Newton A.C., Schreckenberg K. 2003. Commercialising non-timber forest products: first steps in analysing the factors influencing success. *International Forestry Review* 5(2): 128–137.

Marshall E., Rushton J., Schreckenberg K. 2006. *Practical tools for researching successful NTFP commercialization: a methods manual.*

Martínez-Duro E., Ferrandis P., Herranz J.M. 2009. Factors controlling the regenerative cycle of *Thymus funkii* subsp. *funkii* in a semi-arid gypsum steppe: A seed bank dynamics perspective. *Journal of Arid Environments* 73: 252–259.

Marvier M.A., Smith D.L. 1997. Conservation implications of host use for rare parasitic plants. *Conservation Biology* 11: 839–848.

MEA. 2005. Millennium Ecosystem Assessment. *Ecosystems and Human Well-Being: Current State and Trends.* Island Press, Washington, DC

Mejía K.M. 1983. *Palmeras y el Selvícola Amazónico.* UNMSM. Museo de Historia Natural, Lima.

Mejía K.M. 1988. Utilization of palms in eleven mestizo villages of the Peruvian Amazon (Ucayali river). *Advances in Economic Botany* 6: 130–136.

Mejía K.M. 1992. *Las palmeras en los mercados de Iquitos.* IFEA. Bulletin de l'Institut français d'études andines 21: 755–769.

Mejía K.M., Kahn F. 1996. Biologia, ecologia y utilizacion del irapay (*Lepidocaryum gracile* Martius). IIAP. *Folia Amazonica* 8: 19–28.

Miller C. 2002. Fruit production of the ungurahua palm (*Oenocarpus bataua* subsp. *bataua*, Arecaceae) in an indigenous managed reserve. *Economic Botany* 56: 165–176.

Miller G.D., Gaud W.S. 1989. Composition and variability of desert bighorn sheep diets. *Journal of Wildlife Management* 53: 597–606.

Miranda J.F., Montaño A., Zenteno F., Nina H., Mercado J. 2008. *El Majo (Oenocarpus bataua): Una Alternativa de Biocomercio en Bolivia.* TRÓPICO-PNBS-FAN. Ediciones Trópico, La Paz.

Montúfar R. 2010. *La Palma de Ramos en Ecuador. Historia Natural y Estado de Conservación de Ceroxylon echinulatum en las Estribaciones Andinas Nor-occidentales.* IRD, PUCE, Ecuador.

Montúfar R., Pitman N. 2003. *Phytelephas aequatorialis.* IUCN Red List of Threatened Species. Version 2012.1. URL www.iucnredlist.org (accessed on 05.07.2012).

Montúfar R., Pintaud J.-C. 2006. Variation in species composition, abundance and microhabitat preferences among western Amazonian terra firme palm communities. *Botanical Journal of the Linnean Society* 151: 127–140.

Montúfar R., Brokamp G. 2011. Palmeras aceiteras del Ecuador: estado del arte en la investigación de nuevos recursos oleaginosos provenientes del bosque tropical. *Revista Ecuatoriana de Medicina y Ciencias Biológicas* 32: 93–118.

Montúfar R., Laffargue A., Pintaud J.-C., Hamon S., Avallone S., Dussert S. 2010. *Oenocarpus bataua* Mart. (Arecaceae): Rediscovering a source of high oleic vegetable oil from Amazonia. *Journal of the American Oil Chemists' Society* 87: 167–172.

Moraes-R. M. 1998. *Richness and utilization of palms in Bolivia – some essential criteria for their management.* Pp. 269–278. In: Barthlott W., Winiger M. (eds.). Biodiversity: A challenge for development research and policy. Springer, Berlin.

Moraes-R. M. 2004. *Flora de Palmeras de Bolivia.* Herbario Nacional de Bolivia, Instituto de Ecología, Carrera de Biología. Universidad Mayor de San Andrés, La Paz.

Moraes-R. M., Sarmiento J., Oviedo E. 1995. Richness and uses in a diverse palm site in Bolivia. *Biodiversity and Conservation* 4: 719–727.

Mora-Urpí J. 1979. *Pejibaye consideraciones sobre algunos proyectos en marcha.* Asociación Bananera Nacional (Costa Rica) 3: 5–6.

Mora-Urpí J., Weber J.C., Clement C.R. 1997. *Peach palm. Bactris gasipaes* Kunth. Promoting the Conservation and Use of Underutilized and Neglected Crops. 20. Institute of Plant Genetics and Crop Plant Research, Gatersleben/International Plant Genetic Resources Institute, Rome, Italy.

Muñiz-Miret N., Vamos R., Hiraoka M., Montagnini F., Mendelson R.O. 1996. The economic value of managing the asaí palm (*Euterpe oleracea* Mart.) in the floodplains of the Amazon estuary, Pará, Brazil. *Forest Ecology and Management* 87: 163–173.

Murrieta R.S., Dufour D.L., Siqueira A.D. 1999. Food consumption and subsistence in three caboclo populations on Marajoâ Island, Amazonia, Brazil. *Human Ecology* 27: 455–475.

Musselmann L.J. 1996. Parasitic weeds in the southern United States. *Castanea* 61: 271–292.

Musselman L.J. 1975. Parasitism and haustorial structure in *Krameria lanceolata* (Krameriaceae). A preliminary study. *Phytomorphology* 25: 416–422.

Musselman L.J., Dickison W.C. 1975. Structure and development of the haustorium in parasitic Scrophulariaceae. *The Botanical Journal of the Linnean Society* 70: 183–212.

Musselmann L.J., Mann Jr. W.F. 1977. Seed germination and seedlings of *Krameria lanceolata* (Krameriaceae). *Sida* 7: 224–225.

Mutke J., Barthlott W. 2005. Patterns of vascular plant diversity at continental to global scales. *Biologiske Skrifter* 55: 521–531. ISSN 0366-3612.ISBN 87-7304-304-4.

Navarro B. 2006. *Estudio de las Cadenas Productivas de Aguaje y Tagua.* Reserva Nacional Pacaya Samiria, Loreto – Perú. ProNaturaleza, The Nature Conservancy, Agencia de los Estados Unidos para el Desarrollo Internacional (USAID), Lima. Peru.

Navarro J.A. 2009. *Impacto de la Cosecha de Hojas sobre una Población de la Palma Clonal Caraná (Lepidocaryum tenue) en la Estación Biológica El Zafire, Municipio de Leticia, Amazonas (Colombia).* Tesis de Magíster. Universidad Nacional de Colombia, Bogotá.

Navarro J.A., Galeano G., Bernal R. 2011. Impact of leaf harvest on populations of *Lepidocaryum tenue*, an Amazonian understory palm used for thatching. *Tropical Conservation Science* 4: 25–38.

Nepstad D.C., Schwartzman S. (eds.) 1992. Non-timber products from tropical forests: Evaluation of a conservation and development strategy. *Advances in Economic Botany* 9: 1–164.

Neumann R.P., Hirsch, E. 2000. *Commercialisation of Non-timber Forest Products: Review and Analysis of Research.* Bogor: CIFOR, Indonesia.

Newcome J., Provins A., Johns H., Ozdemiroglu E., Ghazoul J., Burgess D., Turner K. 2005. *The Economic, Social and Ecological Value of Ecosystem Services: A Literature Review. Final report for the Department for Environment, Food and Rural Affairs.* Economics for the Environment Consultancy (eftec), London, UK.

Nilsson J.A., Johnson C.D. 1993. A taxonomic revision of the palm bruchids (Pachymerini) and a description of the world genera of Pachymerinae (Coleoptera: Bruchidae). *Memoirs of the American Entomological Society* 41: 1–104.

Noy-Meir I. 1973. Desert ecosystems: environment and producers. *Annual Review of Ecology and Systematics* 4: 25–51.

Núñez L., McRostie V., Cartajena I. 2009. Consideraciones sobre la recolección vegetal y la horticultura durante el formativo temprano en el sureste de la cuenca de Atacama. *Darwiniana* 47: 56–75.

Oboh F.O.J. 2009. The food potential of Tucum (*Astrocaryum vulgare*) fruit pulp. *International Journal of Biomedical and Health Sciences* 5: 57–64.

Ojea E., Martin-Ortega J., Chiabai A. 2010. *Classifying Ecosystem Services for Economic Valuation: the case of forest water services.* BIOECON Conference, Venice 27–28 September 2010.

Orihuela-Ardaya E.F. 2009. Programma Manejo de Bosques de la Amazonia Boliviana: *PFNM en Seis Comunidades Campesinas del Norte Amazónico Boliviano: Causas de Éxito o Fracaso de Comercialización.* PROMAB, Bolivia.

Ortiz Camargo S. 2007. *Potenzial von Märkten für Waldprodukte von Kleinbauern.* Ein Fallbeispiel aus Riberalta, Bolivien. Unpublished MSc-Thesis. Waldbau-Institut. Arbeitsbereich Waldwirtschaft in den Tropen und Subtropen, Albert-Ludwigs-Universität Freiburg.

Pacheco P. 2012. *Soybean and oil palm expansion in South America: A review of main trends and implications.* Working Paper 90. CIFOR, Bogor, Indonesia.

Pacheco-Palencia L.A., Mertens-Talcott S., Talcott S. 2008. Chemical composition, antioxidant properties, and thermal stability of a phytochemical enriched oil from Acai (*Euterpe oleracea* Mart.). *Journal of Agricultural and Food Chemistry* 56: 4631–4636.

Padoch C. 1987. The economic importance and marketing of forest and fallow products in the Iquitos region. *Advances in Economic Botany* 5: 74–89.

Paniagua-Zambrana N.Y., Byg A., Svenning J.C., Moraes M., Grandez C., Balslev H. 2007. Diversity of palm uses in the Western Amazon. *Biodiversity Conservation* 16: 2771–2787.

Parmesan C., Yohe G. 2003. A globally coherent fingerprint of climate change impacts across natural systems. *Nature* 421, 37–42.

Parthasarathy M.V., Klotz L.H. 1976. Palm "Wood". I. Anatomical aspects. *Wood Science and Technology* 10: 215–229.

Pasztor J, Schroeder F. 2012. *A future worth choosing.* In: The road to Rio +20. For a development-led green economy pp. 13–17. Nuñez E.E., (ed.), United Nations Conference On Trade And Development (UNCTAD), New York and Geneva, 71 pp.

Paymal N., Sosa C. 1993. *Amazon worlds: Peoples and cultures of Ecuador's Amazon region*. Sinchi Sacha Foundation, Quito.

Peña-Claros M. 1996. *Producción de Palmito: Manejo Sostenible de Euterpe precatoria (asaí) en la Concesión de Tarumá, Santa Cruz, Bolivia*. Documento Técnico 30. Proyecto BOLFOR, Calle Prolongación Beni 149, Santa Cruz, Bolivia.

Penn J.W. 2008. Non-timber forest products in Peruvian Amazonia: Changing patterns of economic exploitation. *Focus on Geography* 51: 18–25.

Penn J.W., Neise G. 2004. Aguaje palm agroforesty in the Peruvian Amazon. *The Palmeteer* 24: 85–101.

Pennings S.C., Callaway R.M. 1996. Impact of a parasitic plant on the structure and dynamics of salt marsh vegetation. *Ecology* 77: 1410–1419.

Pérez-Arbeláez E. 1956. *Plantas Útiles de Colombia*. Sucesores de Rivadeneyra, Madrid. Pp 831.

Pesce C. 1985. *Oil palms and other oilseeds, of the Amazon*. Reference Publications, Michigan. Pp 28–90.

Peters C.M., Gentry A.H., Mendelsohn R.O. 1989. Valuation of an Amazonian rain forest. *Nature* 339: 655–656.

Peters C.M., Balick M.J., Kahn F. 1989. Oligarchic forests and economic plants in Amazonia: Utilization and conservation of an important tropical resource. *Conservation Biology* 3: 341–349.

Pinedo-Vasquez M., Zarin D., Jipp P., Chota-Inuma J. 1990. Use-values of tree species in a communal forest reserve in northeast Peru. *Conservation Biology* 4: 405–416.

Pintaud J.-C., Anthelme F. 2008. *Ceroxylon echinulatum* in an agroforestry system of northern Peru. *Palms* 52: 96–102.

Pintaud J.-C., Galeano G., Balslev H., Bernal R., Borchsenius F., Ferreira E., de Granville J.-J., Mejía K., Millán B., Moraes M., Noblick L., Stauffer F.W., Kahn F. 2008. Las palmeras de América del Sur: Diversidad, distribución e historia evolutiva. *Revista Peruana de Biología* 15(1): 7–29.

Pimentel D., Pimentel M.H. 2008. *Food, energy, and society*. 3rd edition, CRC, Boca Raton, FL.

Plotkin M., Famolare L. (eds.) 1992. *Sustainable Harvest and Marketing of Rain Forest Products.* Conservation International-Island Press, Washington, DC, pp. 13–15.

Prance G.T. 1979. Notes on the vegetation of Amazonia. III. The terminology of Amazonian forest types subject to inundation. *Brittonia* 31(1): 26–38.

Press M.C., Phoenix G.K. 2005. Impacts of parasitic plants on natural communities. *New Phytologist* 166: 737–751.

Putz F.E. 1979. Biology and human use of *Leopoldinia piassaba. Principes* 23: 149–156.

Pülschen L. 2000. Biology and use of vegetable ivory palms (*Phytelephas* spp.). *Acta Horticulturae* 531: 219–222.

Pyakurel D., Baniya A. 2011. *NTFPs: Impetus for Conservation and Livelihood support in Nepal.* A Reference Book on Ecology, Conservation, Product Development and Economic Analysis of Selected NTFPs of Langtang Area in the Sacred Himalayan Landscape. WWF, Nepal.

Pyhälä A., Brown K., Adger W.N. 2006. Implications of livelihood dependence on non-timber products in Peruvian Amazonia. *Ecosystems* 9: 1328–1341.

Rannow S., Loibl W., Greiving S., Gruehn D., Meyer B. 2010. Potential impacts of climate change in Germany – identifying regional priorities for adaptation activities in spatial planning. *Landscape and Urban Planning* 98: 160–171.

Rautenstrauch K.R., Krausman P.R., Whiting F.M., Brown W.H. 1988. Nutritional quality of desert mule deer forage in King Valley, Arizona. *Desert Plants* 8: 172–174.

R Development Core Team. 2008. *R: A language and environment for statistical 385 computing.* R Foundation for Statistical Computing, Vienna, Austria. ISBN 3-386 900051-07-0. URL http://www.r-project.org. (accessed 05.03.13).

Ren J., Tao L., Liuz X.-M. 2002. Effect of sand burial depth on seed germination and seedling emergence of *Calligonum* L. species. *Journal of Arid Environments* 51: 603–611.

Rogez H. 2000. *Açaí, prepare, composição e melhoramento de conservação.* Editora da Universidade Federal do Pará, Belem. 313 pp.

Rojas-R. R., Salazar-J. C.F., Llerena-F C., Rengifo-S C., Ojanama-V J., Muñoz-I. V., Luque-S. H., Solignac-R. J., Torres-N.D., De Panduro-R. F.M. 2001. Industrialización primaria del aguaje (*Mauritia flexuosa* L.f.) en Iquitos (Perú). *Folia Amazónica* 12: 107–121.

Root T.L., Price J.T., Hall K.R., Schneider S.H., Rosenzweig C., Pounds J.A. 2003. Finger-prints of global warming on wild animals and plants. *Nature* 421: 57–60.

Rudel T., Roper J. 1997. The Paths to Rain Forest Destruction: Cross national Patterns of Tropical Deforestation, 1975–90. *World Development* 25(1): 53–65.

Ruiz H. 1797. Memoria sobre la ratánhia. *Real Academia Médica de Madrid* 1: 349–366.

Ruiz H. 1799. Disertación sobre la raiz de la ratánhia específico singular contra los fluxos de sangre. *Memorias Academia Médica de Madrid* 1: 1–47.

Ruiz-M. J. 1991. *El Aguaje, Alimento del Bosque Amazónico*. PCDEF, Proyecto de Capaci-tación, Extensión y Divulgación Forestal, Pucallpa (Peru). Temas Forestales 8.

Reyes-García V., Huanca T., Vadez V., Leonard W., Wilkie D. 2006. Cultural, practical, and economic value of wild plants: A quantitative study in the Bolivian Amazon. *Eco-nomic Botany* 60(1): 62–74.

Sabbe S., Verbeke W., Van Damme P. 2009a. Analysing the market environment for açaí (*Euterpe oleracea* Mart.) juices in Europe. *Fruits* 64: 273–284.

Sabbe S., Verbeke W., Deliza R., Matta V.M., Van Damme P. 2009b. Consumer liking of fruit juices with different açaí (*Euterpe oleracea* Mart.) concentrations. *Journal of Food Science* 74: 171–176.

Salazar Villón M.B. 2006. *Evaluación de pinturas arquitectónicas de tipo látex con fibras naturales de tagua y cabuya*. Graduation thesis. Escuela Superior Politécnica del Lito-ral, Facultad de Ingeniería en Mecánica y Ciencias de la Producción, Guayaquil, Ec-uador.

Santos L.M.P. 2005. Nutritional and ecological aspects of buriti or aguaje (*Mauritia flexuosa* L. f.): A carotene-rich palm fruit from Latin America. *Ecology of Food and Nutrition* 44: 345–358.

Scariot A. 2001.Weedy and secondary palm species in central amazonian forest fragements. *Acta Botanica Brasilica* 15: 271–280.

Schauss A.G., Wu X., Prior R.L., Ou B., Patel D., Huang D., Kababick J.P. 2006. Phyto-chemical and nutrient composition of the freeze-dried Amazonian palm berry, *Euterpe oleracea* Mart. (Acai). *Journal of Agricultural and Food Chemistry* 54: 8598–8603.

Schippmann U., Leaman D., Cunningham A.B. 2006. *Cultivation and wild collection of medicinal and aromatic plants under sustainability aspects*. In: Bogers R.J., Craker L.E., Lange D. (eds.). Medicinal and Aromatic Plants. Springer, Dordrecht. Wageningen UR Frontis Series no. 17.

Schlüter U.-B., Furch B., Joly C.A. 1993. Physiological and anatomical adaptations by young *Astrocaryum jauari* Mart. (Arecaceae) in periodically inundated biotopes of Central Amazonia. *Biotropica* 25: 384–396.

Scholz E., Rimpler H. 1989. Proanthocyanidins from *Krameria triandra* roots. *Planta Medica* 55: 379–384.

Schreckenberg K., Rushton J., Te Velde D.W. 2006. *NTFP Value Chains: What Happens Between Production and Consumption?* Pp. 97–106 In: Marshall E., Schreckenberg K., Newton A.C. (eds.). Commercialization of Non-timber Forest Products: Factors Influencing Success. Lessons Learned from Mexico and Bolivia and Policy Implications for Decision-makers. UNEP World Conservation Monitoring Centre, Cambridge, UK

Schwarzer C., Cáceres-H. F., Cano A., La Torre M.I.,Weigend M. 2010. 400 years for long-distance dispersal and divergence in the northern Atacama desert – insights from the Huaynaputina pumice slopes of Moquegua, Peru. *Journal of Arid Environments* 74: 1540–1551.

Seeram N.P., Aviram M., Zhang Y., Henning S.M., Feng L., Dreher M., Heber D. 2008. Comparison of antioxidant potency of commonly consumed polyphenol-rich beverages in the United States. *Journal of Agricultural and Food Chemistry* 56: 1415–1422.

SENAMHI. 2013. *Datos Históricos*. Servicio Nacional de Meteorología e Hidrología, Ministerio del Ambiente, Peru. URL http://www.senamhi.gob.pe/main_mapa.php?t=dHi (accessed 20.03.2013).

Shanley P., Pierce A.R., Laird S., Guillen A. (eds.). 2002. *Tapping the Green Market: Certification and Management of Non-timber Forest Products*. People and Plants Conservation Series. London: Earthscan Publications Limited.

Siegel S., Castellan Jr. N.J. 1988. *Non Parametric Statistics for the Behavioral Science*. McGraw-Hill Book Company, Singapore.

Sierra R. 1999. *Propuesta Preliminar de un Sistema de Clasificación de Vegetatición para el Ecuador Continental*. Proyecto INEFAN/GEF-BIRG y Ecociencia, Quito, Ecuador. 194 pp.

Silva S.A.S., De Castro J.C.M., Da Silva T.G., Da-Cunha E.V.L., Barbosa-Filho J.M., Da Silva M.S. 2001. Kramentosan, a new trinorlignan from the roots of *Krameria tomentosa*. *Natural Product Letters* 15: 323–329.

Simpson B.B. 1982. *Krameria* (Krameriaceae) flowers: orientation and elaiophore morphology. *Taxon* 31: 517–528.

Simpson B.B. 1989a. *Krameriaceae*. Flora Neotropica Monograph 49: 1–109.

Simpson B.B. 1989b. Pollination biology and taxonomy of *Dinemandra* and *Dinemagonum* (Malpighiaceae). *Systematic Botany* 14: 408–426.

Simpson B.B. 1991. The past and present uses of rhatany (*Krameria*, Krameriaceae). *Economic Botany* 45: 397–409.

Simpson B.B. 2007. *Krameriaceae*, In: Kubitzki, K., (ed.), The Families and Genera of Vascular Plants IX, 208–212.

Simpson B.B., Neff J.L., Seigler D. 1977. *Krameria*, free fatty acids and oil-collecting bees. *Nature* 267: 150–151.

Simpson B.B., Weeks A., Helfgott D.M., Larkin L.L. 2004. Species relationships in *Krameria* (Krameriaceae) based on ITS sequences and morphology: implications for character utility and biogeography. *Systematic Botany* 29: 97–108.

Simpson P.G., Fineran B.A. 1970. Structure and development of the haustorium in Midasalicifolia. *Phytomorphology* 20: 236–248.

SMUL. 2008. Sächsisches Staatsministerium für Umwelt und Landwirtschaft. *Sachsen im Klimawandel – Eine Analyse*. Dresden.

Soler-Alarcón J.G., Luna-Peixoto A. 2008. Use of terra firme forest by Caicubi Caboclos, middle Rio Negro, Amazonas, Brazil. A quantitative study. *Economic Botany* 62: 60–73.

Soltis D.E., Soltis P.S., Chase M.W., Mort M.E., Albach D.C., Zanis M., Savolainen V., Hahn W.H., Hoot S.B., Fay M.F., Axtell M., Swensen S.M., Prince L.M., Kress W.J., Nixon K.C., Farris J.S. 2000. Angiosperm phylogeny inferred from 18S rDNA, rbcL, and atpB sequences. *The Botanical Journal of the Linnean Society* 133: 381–461.

Sosnowska J., Balslev H. 2009. American palm ethnomedicine: A meta-analysis. *Journal of Ethnobiology and Ethnomedicine* 5: 43.

Spruce R. 1860. On *Leopoldinia piassaba* Wallace. *Journal of the Proceedings of the Royal Society (Botany)* 4: 58–63.

Stagegaard J., Sørensen M., Kvist L.P. 2002. Estimations of the importance of plant resources extracted by inhabitants of the Peruvian Amazon flood plains. *Perspectives in Plant Ecology, Evolution and Systematics* 5(2): 103–122.

Stoian D. 2000. Shifts in forest product extraction: The post-rubber era in the Bolivian Amazon. *International Tree Crops Journal* 10: 277–297.

Stoian D. 2000b. *Variations and Dynamics of Extractive Economies in the Rural-urban Nexus of Non-Timber Forest use in the Bolivian Amazon.* Inaugural-Dissertation zur Erlangung der Doktorwürde der Forstwissenschaftlichen Fakultät der Albert-Ludwigs-Universität Freiburg i. Breisgau.

Stoian D. 2004. *What goes up must come down: The economy of palm heart (Euterpe precatoria Mart.) in the northern Bolivian Amazon.* Pp 111–134. In: Alexiades M.N., Shanley P. (eds.), Forest products, livelihoods and conservation – Case studies of non-timber forest product systems 3 – Latin America.

Stoian D. 2005. Making the Best of Two Worlds: Rural and Peri-Urban Livelihood Options Sustained by Non-timber Forest Products from the Bolivian Amazon. *World Development* 33(9): 1473–1490.

Sunderlin W.D., Angelsen A., Belcher B., Burgers P., Nasi R., Santoso L., Wunder S. 2005. Livelihoods, forests, and conservation in developing countries: An Overview. *World Development* 33(9): 1383–1402.

Svenning J.-C., Randel B. 2013. Disequilibrium vegetation dynamics under future climate change. *American Journal of Botany* 100(7): tba. Published online before print 10 June 2013, doi: 10.3732/ajb.12004692013.

Szaro R.C., Belfit S.C. 1986. Herpetofaunal use of a desert riparian island and its adjacent scrub habitat. *Journal of Wildlife Management* 50: 752–761.

Szaro R.C., Belfit S.C. 1987. *Small Mammal Use of a Desert Riparian Island and Its Adjacent Scrub Habitat.* Fort Collins, CO: U.S..Department of Agriculture, Forest Service, Rocky Mountain Forest and Range Experiment Station. Research Note RM-473, 1–5.

Thomas C.D., Cameron A., Green R.E., Bakkenes M., Beaumont L.J., Collingham Y.C., Erasmus B.F.N., Ferreira de Siqueira M., Grainger A., Hannah L., Hughes L., Huntley B., van Jaarsveld A.S., Midgley G.F., Miles L., Ortega-Huerta M.A., Townsend Peterson A., Phillips O.L., Williams S.E. 2004. Extinction risk from climate change. *Nature* 427: 145–148.

Thomas E., Vandebroek I., Sanca S., Van Damme P. 2009a. Cultural significance of medicinal plant families and species among Quechua farmers in Apillapampa, Bolivia. *Journal of Ethnopharmacology* 122: 60–67.

Thomas E., Vandebroek I., Van Damme P., Goetghebeur P., Douterlungne D., Sanca S., Arrazola S. 2009b. The relation between accessibility, diversity, and indigenous valuation of vegetation in the Bolivian Andes. *Journal of Arid Environments* 73: 854–861.

Tiemann P., Toelg M., Ramos M.H.F. 2007. Administration of Ratanhia-based herbal oral care products for the prophylaxis of oral mucositis in cancer chemotherapy patients: a clinical trial. *Evidence-Based Complementary and Alternative Medicine* 4: 361–366.

Timell T.E. 1957. Vegetable ivory as a source of a mannan polysaccharide. *Canadian Journal of Chemistry* 35: 333–338.

Tinker P.B., Ingram J.S.I., Struwe S. 1996. Effects of slash-and-burn agriculture and deforestation on climate change. *Agriculture, Ecosystems and Environment* 58: 13–22.

Tomlinson P.B. 2006. The uniqueness of palms. *Botanical Journal of the Linnean Society* 151(1): 5–14.

Torres–R. M.C. 2007. *Protocolo de Aprovechamiento in situ de la Especie de Uso Artesanal Werregue (Astrocaryum standleyanum)*. Ministerio de Ambiente, Vivienda y Desarrollo territorial – MAVDT, Instituto de investigación de Recursos Biológicos Alexander Von Humboldt, Instituto de Investigaciones Ambientales del Pacifico.

Torres–R. M.C., Avendano–R. J.R. 2009. *Cartilla de Artesanía en Werregue (Astrocaryum standleyanum) y Chocolatillo (Ischnosiphon arouma) de los Nonam en Puerto Pizario, Bajo río San Juan*. Corporacion Autonoma Regional Del Valle Del Cauca Cvc, Colombia. 29 pp.

United Nations. 2012. United Nations Secretary-General's High-Level Panel on Global Sustainability. *Resilient people, resilient planet: A future worth choosing, Overview*. New York.

Urness P.J. 1973. *Part II: chemical analyses and in vitro digestibility of seasonal deer forages*. In: Deer Nutrition in Arizona Chaparral and Desert Habitats. Arizona Game and Fish Department, Special Report 3: 39–52.

Urness P.J., McCulloch C.Y. 1973. *Part III: nutritional value of seasonal deer diets*. In: Deer nutrition in Arizona chaparral and desert habitats. Arizona Game and Fish Department, Special Report 3: 53–68.

Vallejo M.I., Valderrama N., Galeano G., Bernal R. 2010. *Current status of Euterpe oleracea, the source of palm hearts from the Pacific coast of Colombia*. Poster: Palms 2010, International Symposium on the Biology of the Palm Family. Montpellier, France. (available online: http://fp7-palms.org/component/content/article/18/188-posters-presented-by-fp7-palms-participants.html – accessed 20.11.2010)

Vallejo M.I., Valderrama N., Bernal R., Galeano G., Arteaga G., Leal C. 2011. Current status and perspectives of palm heart production from *Euterpe oleracea* Mart. (Arecaceae), at the Pacific Coast of Colombia. *Colombia Forestal* 14(2): 191–212.

Vander Wall S.B. 1993. A model of caching depth: Implications for scatter hoarders and plant dispersal. *The American Naturalist* 141: 217–232.

Vargas Paredes V., Stauffer F.W., Pintaud J.-C. 2012. Riqueza, usos y conservación de palmas (Arecaceae) en la Reserva Nacional Allpahuayo-Mishana (Perú). *Acta Botánica Venezuelica* 35(1): 53–70.

Vasquez R., Gentry A.H. 1989. Use and misuse of forest-harvested fruits in the Iquitos area. *Conservation Biology* 3: 350–361.

Vásquez-Ocmín P.G., Freitas-Alvarado L., Sotero-Solís V., Paván-Torres R., Mancini-Filho J. 2010. Chemical characterization and oxidative stability of the oils from three morphotypes of *Mauritia flexuosa* L.f., from the Peruvian Amazon. *Grasas y aceites* 61: 390–397.

Velarde M.J., Moraes-R. M. 2008. Densidad de individuos adultos y producción de frutos del asaí (*Euterpe precatoria*, Arecaceae) en Riberalta, Bolivia. *Ecología en Bolivia* 43: 99–110.

Velásquez Runk J. 1995. *Integrating conservation and development: ecological impacts of tagua nut extraction in Comuna Rio Santiago-Cayapas, Ecuador*. Thesis. School of the Environment, Duke University, NC, USA.

Velásquez Runk J. 1998. Productivity and sustainability of a vegetable ivory palm (*Phytelephas aequatorialis*, Arecaceae) under three management regimes in northwestern Ecuador. *Economic Botany* 52(2): 168–182.

Velásquez Runk J. 2001. Wounaan and Emberá use and management of the fiber palm *Astrocaryum standleyanum* (Arecaceae) for basketry in eastern Panamá. *Economic Botany* 55: 72–82.

Villachica H. 1997. *Investigación y desarrollo de sistemas sustentables para frutales nativos Amazónicos, el caso pijuayo*. In: Toledo J.M. (ed.), Biodiversidad y Desarrollo Sostenible de la Amazonía en una Economía de Mercado. Memoria del Seminario-Taller realizado en Pucallpa, Perú. Stansa, Lima.

Vormisto J. 2002. Making and marketing chambira hammocks and bags in the village of Brillo Nuevo, northeastern Peru. *Economic Botany* 56: 27–40.

Vormisto J., Svenning J.-C., Hall P., Balslev H. 2004. Diversity and dominance in palm (Arecaceae) communities in terra firme forests in the western Amazon basin. *Journal of Ecology* 92: 577–588.

Waitman B.A., Vander Wall S.B., Esque T.C. 2012. Seed dispersal and seed fate in Joshua tree (*Yucca brevifolia*). *Journal of Arid Environments* 81: 1–8.

Walter S. 1998. The Utilization of Non Timber Forest Products in the Rainforests of Madagaskar: A Case Study. *Plant Research and Development* 47/48: 121–144.

Warren L.A. 2008. *Demographic Effects of Thatch Harvest and Implications for Sustainable Use of Irapai Palm (Lepidocaryum tenue, Mart.), by Riverine Communities in the Peruvian Amazon*. FIU Electronic Theses and Dissertations. Paper 275. URL http:// digitalcommons.fiu.edu/etd/275 (accessed on 20.10.2011).

Watson D.M. 2009. Parasitic plants as facilitators: more dryad than dracula? *Journal of Ecology* 97: 1151–1159.

Webb R.H., Stielstra S.S. 1979. Sheep grazing effects on Mojave Desert vegetation and soils. *Environmental Management* 3: 517–529.

Weber H.C. 1993. *Parasitismus von Blütenpflanzen*. Wissenschaftliche Buchgesellschaft, Darmstadt.

Weberbauer A. 1945. *El mundo vegetal de los Andes Peruanos (Estudio fitogeográfico)*. Ministerio de Agricultura, Lima.

Weigend M., Dostert N. 2005. Towards a standardization of biological sustainability: Wild-crafting Rhatany (*Krameria lappacea*) in Peru. *Medicinal plant conservation* 11: 24–27.

Weigend M., Rodríguez E.F., Arana C. 2005. The relict forests of northwest Peru and south-west Ecuador. *Revista Peruana de Biología* 12(2): 185–194.

WHO. 2002. WHO *Traditional Medicine Strategy 2002–2005*. World Health Organization, Geneva.

Wickens G.E. 1991. Management issues for development of non-timber forest products. *Unasylva* 42(165): 3–8.

Wild R.G., Mutebi J. 1996. *Conservation through community use of plant resources.* People and plants working paper no. 5. Division of Ecological Sciences, UNESCO, Paris.

Wilson T.B., Witkowski E.T.F. 1998. Water requirements for germination and early seedling establishment in four African savanna woody plant species. *Journal of Arid Environments* 38: 541–550.

Wollenberg E., Inglés A. 1998. *Incomes from the forest: Methods for the development and conservation of forest products for local communities.* CIFOR, Bogor. Pp 227.

Yin R.K. 2003. *Conducting case studies: collecting the evidence.* Pp. 83–106. In: Yin, R.K. (ed.) Case Study Research: Design and Methods. 3rd edition. Applied Social Research Methods Series 5.

Zuidema P.A. 2000. *Demography of exploited tree species in the Bolivian Amazon.* Disserta-tion. ProefschriftUniversiteit Utrecht.

Zuidema P.A., Boot R.G.A. 2000. *Demographic constraints to sustainable palm heart ex-traction from a sub-canopy palm in Bolivia.* Pp. 53–79. In: Zuidema P.A. (ed.), De-mography of exploited Tree species in the Bolivian Amazon. Universiteit Utrecht e Programa de Manejo de bosques de la Amazonia Boliviana. Riberalta, Bolivia.

Internet Resources Consulted

(arranged alphabetically by hyperlink)

http://www.100amazonia.com (accessed 10.11.2010) Strategic consulting, renewable Amazon forest resources. Publisher: 100% Amazonia Exp e Rep LTDA (company), Brazil & United States.

http://www.acai-mania.com (accessed 20.11.2010). Press release on açaí products. Publisher: Açaí-Mania (company), Brazil.

http://www.aguajeperuano.blogspot.com (accessed 05.09.2010). Publisher: Agroindustrial-Floris SAC, "Santa Natura" (company), Peru.

http://www.ainy.fr. (accessed 20.11.2010). Skin care products. Publisher: Aïny (company), France.

http://www.alibaba.com (accessed 20.11.2010). Publisher: Alibaba (company), Hong Kong.

http://www.aromandina.com (accessed 10.11.2010). Essential oils and aroma therapy products. Publisher: Aromandina (company) United States.

http://www.calidris28.com (accessed 12.11.2010). Press release, Schwarze/Weiße Dose 28. Publisher: Calidris28 (company), Luxembourg.

http://www.camdengrey.com (accessed 10.11.2010) Press release. Publisher: Camden-Grey Essential Oils Inc. (company), United States.

http://www.biocomerciosostenible.com/Boletin2.html (accessed 17.10.2010). La Palma Murrapo, Naidio o Açai (Euterpe oleracea). Publisher: Corporación Biocomercio Sostenible, Peru.

http://www.eza.cc (accessed 20.11.2010). Publisher: EZA-Entwicklungszusammenarbeit mit der Dritten Welt GmbH (organisation), Germany.

http://www.indelcusi.com (accessed 20.11.2010). Publisher: Indelcusi (company), Bolivia.

http://www.intracen.org (accessed 20.11.2010). Publisher: ITC, Joint agency of the World Trade Organization and the United Nations.

Appendix A to Chapter 2

Appendix A: Standardized Research Protocol (SRP) - protocol for the implementation of standardized interviews on trade with palm resources

A1 Manual

A1.1 General instructions

For every interview (i.e., interviewee) one master sheet has to be completed and only one consecutive interview number has to be assigned, however, depending on the number of produced/commercialised palm products by individual interviewees one or several annexes have to be completed!

For every data entry (line) in the table "Breakdown of products, species, activities and corresponding annexes" at the end of the master sheet, one annex has to be completed! E.g., if two different products are commercialised deriving from the same palm species (e.g, whole fruit and mesocarp oil), two annexes have to be completed!

Gathering numerical values without information on correct units of weight or time, or, e.g., currency, may be worthless, so please always include these data when completing the forms!

Always start an interview with the master sheet, i.e., with the information on interviewer and interviewee (H01 & H02, and possibly also H03-H06 may be completed before starting the interview in order to save time)!

Always perform the interviews in the default order of this questionnaire using the pre-formulated questions here provided!

Individual interviewers should number their interviews consecutively (e.g., interview 1 performed by N. Valderrama = Valderrama 01, interview with the second interviewee = Valderrama 02) in order to avoid any possibility of confusion (especially in case of several collaborating interviewers working simultaneously).

Fields to be completed in the forms and all questions of the questionnaire are numbered (H01a, H01b, H02a, etc.) to link questions unmistakeably to the corresponding fields where the interviewee's answers have to be completed (shaded in grey)!

When data are transferred to the here provided data capture table (DCT) every (sub)product is filled in a separate line (i.e., one interview may result in several lines in the DCT).

In A13a-c data needs to be separated (using 1, 2, 3, ... or a, b, c, ...) when transferred into the corresponding fields of the DCT.

The product subdivision (I, II, ...) in A14a-d, A15 and A23 is used, when there are slightly different varieties or sizes of a certain type of product (e.g., bracelets elaborated with seeds from different species or elaborated with different number of seeds or differing additional materials) or if by-products derive in the line of production (e.g., press-cake from fruit oil extraction). Only in this case a product subdivision should be done in order to save time and paper sheets.

A 1.2 Questionnaire

Master Sheet
Interviewer
H01a-b   [H01a] Please complete your full name and [H01b] the present date!
H02a-b   [H02a] Please complete your consecutive interview number and [H02b] the institution's name you are affiliated with!

Interviewee and location

H03a-e   Please complete detailed information on the location where the study is conducted: [H03a] country, [H03b] province, [H03c] department, [H03d] municipality, and [H03e] name of city/village/community (or distance and direction to the closest settlement)!
H04a-c   Please complete GPS data of the location (if available): [H04a] latitude, [H04b] longitude, and [H04c] altitude!
H05a-b   Please specify how the interviewee contact was established: Mark [H05a] if self-established or [H05b] if established through mediator! Please provide the mediator's full name (if available) and/or the name of mediating community or company!
H06a   Please complete, if you have information on interviewee's ethnicity (name of ethnic group, or, e.g., "mestizo")!
H06b   Please complete interviewee's gender (mark m for male or f for female)!

H01 & H02, and possibly also H03-H06 may be completed before starting the interview in order to save time!

Please read the following questions (and possible answers where indicated with *) and complete the interviewee´s answers in the corresponding fields of the forms!

H07	"Please, give me your full name!" [explain that names will be kept anonymous]
H08a	"What is your age?" [Complete exact interviewee´s age if imparted!]
H08b*	"Did you (and your family) have always resided here or did you move to this location recently?" Mark [H08b] if interviewee is native to the location or [H08c] if interviewee has moved in recently! In the latter case also ask:
H08c	"For how many years do you live here?" [Complete number of years the interviewee is residentiary in the study location]
H09a-d*	"When working with palms, [H09a] do you work solitary, [H09b] with help of your family or [H09c] do you cooperate with other persons or [H09d] is it that none of these options are correct?" Mark [H09a] if interviewee works solitary! In the second case ask [H09b]: "How many family members (i.e., how many adults and how many children) help you?", or in the third case [H09c]: "With how many persons do you cooperate commonly?", or if none of the given options fit (e.g., if interviewee employs workmen) [H09d] "Please specify with whom you are collaborating and how exactly!" [Complete provided information (number and duration of cooperations or employed workmen!]
H10	"Regarding your work with palms, are you associated with any institution, company, or community?" In case of disapproval mark "no", in case of approval, mark "yes" and ask: "What is the name of the institution/company/community you are associated with?"[complete provided name!]
H11a-c*	"What period of time do you work with palms annually? Mark [H11a] if interviewee works year-round, mark [H11b] and complete number of months if interviewee only works several months per year with palms, or mark [H11c] and specify, if it is more complicated, e.g., the interviewee only works a number of days in certain months with palms!
H12a	"Does the work with palms represents your main occupation?" Mark [H12a] in case of approval or (only in case of disapproval) ask:
H12b	"What other occupations are you pursuing? [Complete other occupations!]
H13a-j*	"Regarding your work with palms, please tell me in what activities you are involucrated? [Read the given options (harvester, grower, supplier, carrier, producer, intermediary, salesman, exporter, supporter, inspector) and mark were approved (more than one rôle may be possible here!)!]
H14a-g	For **every** approved activity (rôle) ask: "Do you need a permit to comply with your activity?" [if interviewee acts as a supporter or inspector (i.e., NGO staff or government official) ask: "Do you know, whether a permit is required for harvest, cultivation, supply, transport, production, trade (national) or export of palm resources?" In case of disapproval mark "no", in case of approval mark "yes" in the field corresponding to the interviewee´s activity and ask: "What type of permit is required?" (and specify! E.g., collection permit, companies register, certificate of sanitary conditions, export permit, ...) [if interviewee acts as a supporter or inspector,

	ask: "What type of permit is required for harvest, cultivation, supply, transport, production, trade (national) and export of palm resources?"] and also ask:
H15a-g	"How do you obtain the required permit(s)?" (or if interviewee acts as a supporter or inspector, ask: "How are required permits obtained?") [Complete interviewee's answer(s) in the corresponding field(s)!]
H15h-n	In case a permit is required, ask: "Is there a regular control of required permit(s)?" [In case of disapproval mark "no", in case of approval mark "yes" in the field corresponding to the interviewee's activity]
H16a-c	"What is your cash income only through your activity with palms per month (or per year)?" [Complete income (H16a), mark the correct unit of time (H16b) and provide currency infromation (H16c)]
H16d	"What percentage of your overall cash income comes from your activity with palms?" [Complete provided percentage!]
H17a-f	"How many years have you been involved in the (only use applicable options!) harvest/cultivation/supply/transport/production/commercialisation of palms/palm products?" [Complete number of years for every activity the interviewee is involved in!]
H18a-h*	"With whom do you interact directly to comply with your activity?" [Read the given options (landowner, harvester, supplier, carrier, salesman, government officials, NGO staff, other) and mark the applicable. Specify "other" in case the interviewee interacts with stakeholders not mentioned in the here provided list of options!]
H19a	"Have you encountered any problems to comply with your activity?" [Mark "no" in case of disapproval, mark "yes" in case of approval and ask H19b-d]
H19b	"Please describe one by one the (three) most prominent problems as short as possible!" [Complete described problems (1, 2, 3)]
H19c	Ask for every mentioned problem: "What is the cause of that (1, 2, 3) problem?" [Complete described causes]
H19d	Ask for every mentioned problem: "How do you think the problem (1, 2, 3) can be solved? [Complete provided possible solution(s)!]

Breakdown of products, species, activities and corresponding annexes

H20 **"With what palm raw material or palm product do you work and what exactly is your activity in that? [All mentioned palm raw materials/products have to be listed in the here provided table (best fill in together with the interviewee) with corresponding palm species name (if available) or vernacular name and all activities (rôle) the interviewee is involved with!]**

For <u>every commercialised palm product</u> exactly <u>one annex</u> has to be completed!

Annex: product details

Interviewer

A01a-b [A01a] Please complete your full name and [A01b] your consecutive interview number!

A02 Please complete the name of palm product under study!

A03 Please complete the palm species name of which raw material(s) or the product is obtained!

In case the interviewee is not active in harvest or cultivation start with [A11a]!

Raw material/product source

A04 "What plant part is harvested/required for production?" [Complete plant part, e.g., leaf, leaf petiole, infructescence, fruit, fruit kernel!]

A05a "Do you harvest required raw material/plant parts yourself?" [In case of disapproval, continue with A11! In case of approval, ask:]

A05b-c* "Where do you harvest required raw materials/plant parts? [Read the provided options: "Do you harvest from the wild (i.e., forest) or do you harvest from your own cultivation area (e.g., chagra, purma, finca) or do you harvest from the wild as well as from your own cultivation area?" [Mark applicable!]

A06 "How do you harvest the plant material and what kind of tool(s) do you need for that?" [Complete a short description of the applied harvest method with information on required tools! E.g., spear leaves are cut with a saw with long handle and carried home, or the tree is felled with an axe, then the infructescences are separated with a machete and carried home.]

A07a "What amount of raw material do you harvest per month?" [Complete interviewee´s answer with information on unit of weight!]

A07b "What amount of raw material do you harvest per tree?" [Complete interviewee´s answer with information on unit of weight!]

A08a "After what time you can harvest from the same tree again?" [Complete interviewee´s answer (e.g., twice per year, every second year, only once)!]

A08b "What time is needed for the natural production of extracted plant parts?" [E.g., every year the palm produces two new infructescences, i.e. 2/year.]

A08c "What is the area size where you harvest from/cultivate palm species?" [Complete area size with corresponding unit, e.g., in ha or m^2, 1 ha = 10.000 m^2 = 100 m x 100 m! Indicate area size of wild harvest with "W" and area size of cultivation with "C" and separate figures with "/" if both options are applicable, e.g., W50 ha/ C0.5 ha!]

A09a "What is the number of harvestable individuals of the required palm species in the harvest area?" [and/or in case of own cultivation ask:] "What is the number of harvestable palm trees in your cultivation area?" [Complete number of harvestable individuals in the harvest area and/or cultivation area (if both applicable, separate figures with "/", indicate wild harvest with "W", and harvest from cultivation area with "C", e.g., W200/C30!)!]
[In case the interviewee is not cultivating, continue with A11!]

A09b	"How many years after planting the palms can be harvested for the first time?" [Complete number of years!]
A09c	"Since when do you cultivate this palm species?" [Complete number of years!]
A10a-c	"Who is the owner of the land where you cultivate this palm species?" Mark [A10a] if the interviewee is the landowner, mark [A10b] if it is communal territory or [A10c] if the area neither is the interviewee´s property nor communal territory! [In the latter case ask:]
A10d	"Do you have to pay lease for the area?" (In case of disapproval continue with [A11], in case of approval mark "yes" and ask: "How much lease do you pay per month?" [Complete lease per month and specify currency!]
A11a	"Do you (also) buy palm raw material/products?" In case of disapproval continue with [A13], if interviewee performs pre-processing of raw materials/products; or continue with [A14], if interviewee is producing, bottling or repacking raw materials/products; or continue with [A17], if interviewee only buys and resales unprocessed raw materials/products! In case of approval mark with "X" and ask:
A11b	"What amount of raw material/products do you buy per month?" [Complete purchased amount per month with information on unit!]
A11c	"From whom do you buy raw material/products?" [Complete name(s) of salesman or company!]
A11d	"Where from (location) do you buy raw material/products?" [Complete location name(s)!]
A11e	"What is the purchase price of raw material/products?" [Complete purchase price with information on unit and currency!]
A12a	In case interviewee harvests raw material ask: "Do you follow a management plan or do you provide any kind of source documentation?" In case interviewee buys raw materials or products, ask: "Do the raw materials/products come with any kind of source documentation?" In case of approval ask:
A12b	"Is it possible to obtain a copy or foto of the management plan and/or source documentation?"

Pre-processing of raw material(s)/products

A13	"Does the raw material that you use/buy needs to be pre-processed before production?" In case of approval continue with [A13a] and complete the table "Pre-processing of raw material(s)/products" In case of disapproval continue with [A14] if interviewee produces or with [A17] if interviewee only buys and resells raw material(s)/products!

If any kind of pre-processing is realised, ask:

A13a	"What different stages of pre-processing are realised?"
A13b	"How are the mentioned stages of pre-processing realised? Are any additional materials or tools required?"
A13c	"How much time do you need per unit to realise the mentioned stages of pre-processing?"

Products and production

A14   Complete all different (sub-)products and by-products that are or might be commercialized by the interviewee into the table "Products and production"! If there are different sub-products or by-products please assign them to I, II, ... in order to separate production details and information on marketed amounts, market locations and prices!

A14a   "What (sub)-products do you produce? Are any by-products obtained during the production process?" [List all different (sub-)products and by-products!]

For every (sub-)product or by-product listed in [A14a] ask:

A14b   "What amount of raw material is required for the production of one unit of (sub-)product (and by-product)?" [Specify all units!]

A14c   "How much time is required for the production of one (sub-)product unit?

A14d   "What is the usual selling price (minimum-maximum) for every (sub-)product (and by-product)?" [Specify unit and currency!]

A15   "What amount of (sub-)products (and by-products) do you usually produce every month?" (minimum-maximum and/or mean, specify unit!)

A16   "Why do you produce this amount? What are the limiting factors of production?"

A17   "Do you implement any type of batch coding?" In case of disapproval mark "no", in case of approval mark "yes" and ask "Please specify what information is provided by the batch code (e.g., harvest source, date, certain quality, ...)?"

Trade

If there are different sub-products (or by-products from production) commercialised assign them to I, II, ... in order to separate trade details and information on marketed amounts, market locations and prices!

A18   "How are the selling prices determined?" (e.g., quality, variety, season, location, ...)

A19a   "Where do you sell raw materials/products?" [complete location(s)!]

A19b   "What are the transportation costs to the market/business partner?" [Specify transportation costs per unit with information on currency!]

A20   "To whom do you sell raw materials/products?" [please provide name of business partner, company, or intermediary!]

A21   "What is the final use of raw materials/products you sell?" [Specify!]

A22   "What are the limiting factors of sale?" [Specify!]

A23   "What amount of raw materials/each (sub-)product and/or by-product do you sell per month? [Please specify with unit information, using I,II, ..., if there are several raw materials/(sub-)products/by-products commercialised!]

Appendix A2.1 SRP Forms: Master Sheet

Master Sheet		*-please complete this form for every informant!-*		
H01	Interviewer (full name)		Date	
H02	Interv.-No.	Institution		

Interviewee and location

H03	country		province		
	department		municipality		
	name of city/village/community				
H04	GPS data	S/N	E/W	altitude	m
H05	contact	O direct	O mediator (full name):		
H06	ethnic group		gender	O m	O f
H07	interviewee's full name				

H08	age	years	residentiary in interview location	O native	O residentiary since _____ years
H09	O individual, working solitary	O family business with _____ adults and _____ children	O _____ cooperating persons		
	O other (specify):				
H10	associated with institution/company/community	O no O yes:			
H11	time working with palms per year	O perennial	O __ /12 months	O other (specify):	
H12	palm business main occupation?	O yes	O no, part-time, other occupation(s):		

H13	activity (regarding palm business)	O harvester	O grower	O supplier	O carrier	O producer
		O intermediary	O salesman	O exporter	O supporter	O inspector

H14	type of permit required to pursue your activity?	transport	O no O yes:
	harvest	O no O yes:	production	O no O yes:
	cultivation	O no O yes:	national trade	O no O yes:
	supply	O no O yes:	export	O no O yes:

H15	How do you obtain the required permit(s) for ...?		Control (of permit)
	harvest		O no O yes
	cultivation		O no O yes
	supply		O no O yes
	transport		O no O yes
	production		O no O yes
	national trade		O no O yes
	export		O no O yes

H16	cash income		O /month O /year	currency		percentage of tot.?	%

H17	How many years have you been involved in the harvest/cultivation/supply/transport/production/commercialisation of palms/palm products?					
palms, level of experience	harvest	cultivation	supply	transport	production	commercialisation
	years	years	years	years	years	years

H18 With whom do you interact directly to comply with your activity?

O landowner	O harvester	O supplier	O carrier	O salesman	O government officials	O NGO staff
O Other (please specify!):						

H19	Have you encountered any problems to comply with your activity?	O no O yes, specify: description, cause, possible solution	
	description	cause	possible solution
1.			
2.			
3.			

Breakdown of products, species, activities and corresponding annexes						
H20	product/ raw material	palm species (scientific and/or vernacular name)	wild harvest	cultivation	supply/transport	production	commercialisation
Examples:						
palmito	Euterpe oleracea		X	/ X		X
mesocarp oil	ungurahua	X		/ X	X	X
fibra, escoba	Aphandra natalia	X	X	X / X	X	X

Appendix A2.2 SRP Forms: Annex

Annex: product details		*-please complete this sheet for every product!-*	
A01	**Interviewer (full name)**		**Interv.-No.**
A02	**product name**		
A03	**palm species**		

Raw material/product source

A04	plant part(s) required (e.g., leaf)				
A05	raw material source	O own harvest	harvest area	O forest (wild)	O own cultivation area
A06	harvesting method (tools used)				
A07	harvested amount per month		per tree		
A08	harvest frequency (same tree)		nat. production	_____ /year	area size
A09	no. of trees		1. harvest after	_____ year(s)	own cultivation performed since _____ year(s)
A10	land tenure	O landowner	O community	O leased	leasing costs	O yes, _____ per month
A11	O purchased	per month	source (seller)		
	source (location)		purchase price		
A12	management plan or source documentation?	O no O yes	copy or photography available?	O no O yes	

Pre-processing of raw material(s)/products

A13	pre-processing stage (what is done?)	how? (additional raw material or tool(s) required?)	time/unit

Products and production

A14	product (name/description)	raw material required per unit	required working time/unit	price/unit
I				
II				
...				

A15	amount of production per month	I:
A16	why this amount, what are the limiting factors?	
A17	batch coding implemented?	O no O yes, providing information on:

Trade

A18	prices are determined by ... ?	
A19	point-of-sale	transportation costs
A20	to whom do you sell?	
A21	final use of raw material/product	
A22	limiting factors of sales?	
A23	sold amount per month?	I:

189

Appendix A3.1 SRP DCT: Master Sheet

Interviewer				Interviewee and location				
H01	H01a	H02	H02a	H03	H03a	H03b	H03c	H03d
Interviewer (full name)	Date	Interview	Institution	Country	Province	Department	Municipality	City/village/community

Examples

D. Baldassari	04.Julio 2011		1 PUCE	Ecuador				Barrio; Las Tolas - Gualea
D. Baldassari	04.Julio 2011		1 PUCE	Ecuador				Barrio; Las Tolas - Gualea
D. Baldassari	04.Julio 2011		1 PUCE	Ecuador				Barrio; Las Tolas - Gualea
D. Baldassari	04.Julio 2011		2 PUCE	Ecuador				

H04

GPS- coordinates

latitude	longitude	altitude
S 00° 04' 42.4"	W 078° 46' 33.9"	1824 m
S 00° 04' 42.4"	W 078° 46' 33.9"	1824 m
S 00° 04' 42.4"	W 078° 46' 33.9"	1824 m

190

Appendix A3.1 SRP DCT: Master Sheet (continued)

H05	H06	H06a		H07	H08
Contact	Ethnic group	Gender		Interviewee	Age
direct	intermediate	m	f	full name	years
X	-		X	[anonymized]	45
X	-		X	[anonymized]	45
X	-		X	[anonymized]	45

H08a	H08b	H09	
Residentiary in interview location		Cooperator	
O native	residentiary since	working solitary	family with ? adults
X			2
X			2
X			2

Appendix A3.1 SRP DCT: Master Sheet (continued)

H09 (cont.)		H10	H11	
		Associated with	Time working with palms per year	
? cooperating persons	*other*	*institution/company/community*	*? months*	*perennial*
and ? kids		Asociación de artesanos Muyucunayumbo		X
		Asociación de artesanos Muyucunayumbo		X
		Asociación de artesanos Muyucunayumbo		X

H12	H13			
Palms, main occupation?	Activity (palms)			
no, part-time, other occupation	*harvester*	*grower*	*supplier*	*carrier*
yes, palms				
X, Turismo, Ganadero				
X, Turismo, Ganadero				
X, Turismo, Ganadero				

Appendix A3.1 SRP DCT: Master Sheet (continued)

H13 (cont.)

Activity (palms)

producer	intermediary	salesman	exporter	supporter	inspector
X		X			
X		X			
X		X			

H14 y H14a

Permit required to pursue your activity? What type?

harvest	cultivation	supply	transport	production	trade	export
				no	no	
				no	no	
				no	no	

H15

How do you obtain the required permit(s) for ...?

harvest	cultivation	supply	transport

Appendix A3.1 SRP DCT: Master Sheet (continued)

H15 (cont.)			H15a		H16			H16a
			Controlled?		Income (from palms)			Percentage
produce	trade	export	(of permit)		/month	/year	currency	of total income
-	-	-	-		-	150	USD	-
-	-	-	-		-	-	-	-
-	-	-	-		-	-	-	-

H17				
How many years involved in …?				
harvest	cultivation	supply	transport	production	trade
12	12			12	11
-	-			-	-
-	-			-	-

194

Appendix A3.1 SRP DCT: Master Sheet (continued)

With whom do you interact? (H18)								Problems? (H19)	
landowner	harvester	supplier	carrier	salesman	government officials	NGO staff	other	O no	O yes
	X, de [anonymized]	X		X					X

H19a	**H19b**	**H19c**
Problem description	Problem cause	Possible solution
1 2 3 4	1 2 3 4	1 2 3 4
falta de contactos, no hay comercialización	falta de contactos a quien vender	formar una cooperativa y buscar contactos

Appendix A3.1 SRP DCT: Master Sheet (continued)

Desglose de productos, especies y actividades y correspondientes anexos								Contact
			H20					
Product/raw material	Palm species name	Wild harvest	Cultivation	*Supply*	*Transport*	Production	Trade	Mail/Phone
Artesania de Tagua	*Phytelephas aequatorialis*					X	X	
Artesania de palmito	*Prestoea acuminata*	X				X	X	
Artesania de pambil	*Iriartea deltoidea*					X	X	

Appendix A3.2 SRP DCT: Annex

Interviewer, product name, and palm species			
A01	A01a	A02	A03
Interviewer (full name)	Interview	Product name	Palm species

Examples

Daniela Baldassari	1	Artesania de Tagua	*Phytelephas aequatorialis*
Daniela Baldassari	2		

Raw material/product source		
A04	A05	A05a
Plant part/s required	Source of raw material/s	O own cultivation
	O own harvest *O forest (wild)*	
Semilla		

A08a	A08b	A09	A09a
Natural production	Area size	No. of trees	First harvest
			after how many years

A06	A07	A07a	A08
Harvesting method	Harvested amount		Harvest frequency
tools used?	*per month*	*per tree*	*from individual tree*

Appendix A3.2 SRP DCT: Annex (continued)

A09b	A10	A10a	A11	A11a	A11b	A11c	A12
Own cultivation since *in years*	Area of own cultivation *land tenure*	*costs (currency?)*	Purchase of raw material or finished products O *purchased*	*amount/month*	*source and seller*	*price (currency?)*	Management plan/source documentation? *type of document*
			X	se compra 5 qq/año, se -		8 USD/qq (+15 $ transporte)	

Pre-processing of raw material/products

	A13	
Pre-processing stage *what is done?*	Additional raw material or tools *how?*	time/unit
1. secar, 2. corte en tajada, 3. si 1. Horno, 2. sierra electrica, 3. b	1. 2,5-3 dias, 2. 2 min/tagua, 3. 3 Bisoteria	

Products and production

Product	A14		
	Amount of raw material *per unit*	Required working time *per unit*	Sales price *per unit*
Bisuteria	- (depende)	2 semanas/producto	0.8-10 $/producto

Trade

A15	A16	A17	A18
Amount of units produced *per month*	Why this amount? *what are the limiting factors?*	Batch coding implemented? O *no*	O *yes (information?)*	Prices are determined by … ?
depende de pedidas	humedad, demanda	X	calidad, tiempo de elaboración, precios de la competencia

A19	A19a	A20	A21	A22	A23
Point-of-sale *location*	Transportation costs	To whom do you sell? *buyer/company*	Final use?	Limiting factors of sales?	Sold amount per month?
voluntarios, comerciantes en Tulipe y Quito	a Quito 1 $, al exterior el cliente pag voluntarios, intermediarios y turistas		Bisuteria	competencia	depende de pedido

Appendix B to Chapter 6

Appendix B Raw data and results from statistical analyses

B1.1 Raw data table from population inventories (Bal – Balsas, Car – Caraz, Tar – Tarma, Oma – Omate, SAA – San Antonio A, SAB – San Antonio B, Chu – Chuquibamba, detailed information on the study locations may be obtained from Table 1, na – not available)

Study location name & No.	No. of counted individuals per size class			Total No. of individuals	Counted No. of digging holes
	Seedling	Juvenile	Adult		
Bal101	0	0	2	2	2
Bal102	0	1	2	3	2
Bal103	0	0	2	2	3
Bal104	0	0	1	1	7
Bal105	0	0	0	0	2
Bal106	0	0	0	0	1
Bal107	0	0	1	1	3
Bal108	0	1	3	4	1
Bal109	0	0	3	3	0
Bal110	0	0	2	2	1
Bal111	0	0	5	5	0
Bal112	0	1	10	11	0
Bal113	0	0	6	6	1
Bal114	0	0	0	0	0
Bal115	0	4	2	6	0
Bal116	0	0	6	6	0
Bal201	0	1	10	11	0
Bal202	0	1	4	5	0
Bal203	0	1	4	5	0
Bal204	0	0	1	1	0
Bal205	0	0	3	3	0
Bal206	0	1	6	7	0
Bal207	0	0	3	3	0
Bal208	0	0	2	2	0
Bal209	0	1	2	3	0
Bal210	0	0	7	7	0
Bal211	0	0	5	5	0
Bal212	0	0	10	10	0
Bal213	0	0	2	2	0
Bal214	0	0	1	1	0
Bal215	0	5	8	13	0
Car101	2	0	9	11	0
Car102	0	2	7	9	0
Car103	1	0	8	9	0
Car104	0	5	28	33	0

B1.1 Raw data table from population inventories (continued)

Car105	0	2	5	7	0
Car106	0	2	8	10	0
Car107	1	3	15	19	0
Car108	34	10	24	68	0
Car109	14	10	20	44	0
Car110	3	2	17	22	0
Car111	19	0	11	30	0
Car112	18	3	47	68	0
Car113	73	6	37	116	0
Car114	4	3	20	27	0
Car115	0	1	8	9	0
Car201	7	8	33	48	0
Car202	12	10	35	57	0
Car203	0	2	33	35	0
Car204	26	11	62	99	0
Car205	0	3	21	24	0
Car206	7	13	48	68	0
Car207	8	12	32	52	0
Tar101	0	1	6	7	na
Tar102	0	3	1	4	na
Tar103	0	1	4	5	na
Tar104	0	3	5	8	na
Tar105	0	2	5	7	na
Tar106	0	1	4	5	na
Tar107	0	2	3	5	na
Tar108	0	0	0	0	na
Tar109	0	0	1	1	na
Tar110	0	1	3	4	na
Tar111	0	2	2	4	na
Tar112	0	4	2	6	na
Tar113	0	0	2	2	na
Tar114	0	1	7	8	na
Tar115	0	7	6	13	na
Tar116	0	3	4	7	na
Tar117	0	9	9	18	na
Tar201	0	4	10	14	na
Tar202	0	0	2	2	na
Tar203	0	0	1	1	na
Tar204	0	0	6	6	na
Tar205	0	0	5	5	na
Tar206	0	0	3	3	na
Tar207	0	2	12	14	na
Tar208	0	4	3	7	na
Tar209	0	0	7	7	na

B1.1 Raw data table from population inventories (continued)

Tar210	0	6	11	17	na
Tar211	0	3	5	8	na
Tar212	0	5	4	9	na
Tar213	0	4	8	12	na
Tar214	0	4	8	12	na
Oma101	8	3	2	13	na
Oma102	0	0	0	0	na
Oma103	1	0	0	1	na
Oma104	1	0	0	1	na
Oma105	0	0	1	1	na
Oma106	0	0	0	0	na
Oma107	0	0	0	0	na
Oma108	0	0	0	0	na
Oma109	0	0	0	0	na
Oma110	0	0	0	0	na
SAA101	0	0	1	1	0
SAA102	0	0	0	0	0
SAA103	0	0	0	0	0
SAA104	0	0	0	0	0
SAA105	0	0	0	0	0
SAA201	2	5	1	8	2
SAA202	3	3	2	8	2
SAA203	0	0	1	1	0
SAA204	0	1	0	1	0
SAA205	0	2	2	4	0
SAA206	1	3	6	10	1
SAA207	1	1	10	12	3
SAA208	0	2	6	8	1
SAA209	0	5	4	9	2
SAA210	2	3	5	10	3
SAA211	0	0	2	2	0
SAA212	0	3	2	5	0
SAA213	5	7	2	14	2
SAA214	2	11	2	15	6
SAA215	3	3	6	12	5
SAA216	2	4	3	9	3
SAA217	0	2	1	3	3
SAA218	4	1	24	29	4
SAA219	2	3	14	19	3
SAA220	8	4	1	13	1
SAA221	24	20	1	45	2
SAA222	3	7	2	12	3
SAA223	0	0	2	2	0
SAA301	4	6	16	26	0

B1.1 Raw data table from population inventories (continued)

SAA302	0	0	10	10	0
SAA303	2	0	5	7	0
SAA304	1	0	0	1	0
SAA305	0	1	2	3	0
SAA306	0	0	0	0	0
SAA307	0	3	7	10	0
SAA308	3	7	8	18	0
SAA309	1	1	9	12	0
SAA310	1	0	2	3	0
SAA311	1	3	6	10	0
SAA312	0	2	7	9	1
SAB101	0	1	3	4	10
SAB102	0	5	4	9	13
SAB103	1	8	5	14	8
SAB104	2	12	9	23	11
SAB105	1	8	11	20	8
SAB106	0	0	4	4	0
SAB107	0	0	1	1	6
SAB108	0	1	0	1	5
SAB109	0	0	6	6	2
SAB110	0	1	4	5	4
SAB111	0	0	0	0	0
SAB112	0	0	2	2	0
SAB113	0	0	0	0	5
SAB114	0	3	2	5	4
SAB115	0	11	5	16	0
SAB116	0	0	6	6	6
SAB117	0	6	5	11	0
SAB118	0	0	3	3	0
SAB201	0	0	2	2	10
SAB202	2	6	8	16	0
SAB203	4	13	6	23	8
SAB204	6	10	2	18	17
SAB205	0	6	19	25	9
SAB206	0	11	14	25	1
SAB207	0	0	1	1	0
SAB208	0	0	1	1	1
SAB209	2	1	8	11	4
SAB210	1	1	4	6	4
SAB211	0	4	4	8	4
SAB212	0	0	1	1	0
SAB213	0	0	0	0	8
SAB214	0	8	3	11	6
SAB215	0	8	5	13	2

B1.1 Raw data table from population inventories (continued)

SAB215	0	8	5	13	2
SAB216	0	5	7	12	7
SAB217	0	5	9	14	2
SAB218	0	2	3	5	0
SAB301	0	2	0	2	11
SAB302	0	6	3	9	8
SAB303	1	13	4	18	7
SAB304	4	7	5	16	9
SAB305	1	9	17	27	9
SAB306	0	5	7	12	0
SAB307	0	0	0	0	2
SAB308	0	3	2	5	5
SAB309	0	0	2	2	5
SAB310	0	0	0	0	0
SAB311	0	1	3	4	10
SAB312	0	0	0	0	0
SAB313	0	3	2	5	2
SAB314	0	5	6	11	18
SAB315	4	8	3	15	6
SAB316	0	1	4	5	0
SAB317	0	2	8	10	10
SAB318	0	2	2	4	0
Chu101	7	18	6	31	na
Chu102	10	6	4	20	na
Chu103	8	20	4	32	na
Chu104	5	11	5	21	na
Chu105	3	15	6	24	na
Chu106	11	6	6	23	na
Chu107	18	14	4	36	na
Chu108	12	13	10	35	na
Chu109	10	6	12	28	na
Chu110	12	8	5	25	na
Chu201	18	13	2	33	na
Chu202	12	17	2	31	na
Chu203	17	10	5	32	na
Chu204	15	15	6	36	na
Chu205	14	10	3	27	na
Chu206	13	8	2	23	na
Chu207	12	6	2	20	na
Chu208	10	13	4	27	na
Chu209	11	13	3	27	na
Chu210	11	12	1	24	na
Chu301	2	2	14	18	na
Chu302	3	6	13	22	na

B1.1 Raw data table from population inventories (continued)

Chu303	0	6	20	26	na
Chu304	1	5	20	26	na
Chu305	1	7	16	24	na
Chu306	9	6	13	28	na
Chu307	0	5	15	20	na
Chu308	2	2	20	24	na
Chu309	1	4	22	27	na
Chu310	1	8	21	30	na
Chu401	1	15	7	23	na
Chu402	2	17	13	32	na
Chu403	7	10	10	27	na
Chu404	5	15	10	30	na
Chu405	4	10	12	26	na
Chu406	3	8	12	23	na
Chu407	0	6	11	17	na
Chu408	10	13	13	36	na
Chu409	1	13	12	26	na
Chu410	1	12	8	21	na
Chu501	0	2	7	9	na
Chu502	0	2	5	7	na
Chu503	0	1	13	14	na
Chu504	0	8	16	24	na
Chu505	0	0	13	13	na
Chu601	0	2	12	14	na
Chu602	0	6	8	14	na
Chu603	0	4	8	12	na
Chu604	0	7	10	17	na
Chu605	0	2	10	12	na

B1.2 Raw data from the germination experiment

Pot No.	Sow depth and seed treatment					
	0 cm – untreated	5 cm – untreated	10 cm – untreated	0 cm – pre-treated	5 cm – pre-treated	10 cm – pre-treated
1	1	8	9	1	8	8
2	4	7	10	2	9	10
3	2	9	8	0	10	8
4	3	10	9	0	8	10
5	4	7	10	4	8	9
6	3	8	8	3	8	8
7	5	8	9	1	9	8
8	1	8	8	1	9	8
9	4	8	8	2	8	9
10	1	9	9	2	8	8

B2.1 Results from the statistical analyses (L – Levene's test, M – Mann-Whitney-U test) of the pairwise tested population inventory data (Bal – Balsas, Car – Caraz, Tar – Tarma, Oma – Omate, SAA – San Antonio A, SAB – San Antonio B, Chu – Chuquibamba, detailed information on the study locations may be obtained from Table 1, na – not available; results of the Mann-Whitney-U test in brackets in case the Levene's-test was significant at the $p < 0.05$ level).

B2.1.1 Seedlings

	Bal2	Car1	Car2	Tar1	Tar2	Oma1	SAA1	SAA2	SAA3
Bal1	na	L: p = 0.001; (M: W = 40, p = 0.0002)	L: p = 0.001; (M: W = 16, p = 0.0003)	na	na	L: p = 0.009; (M: W = 56, p = 0.0256)	na	L: p = 0.015; (M: W = 72, p = 0.0002)	L: p = 0.000; (M: W = 40, p = 0.0007)
Bal2	-	L: p = 0.001; (M: W = 37.5, p = 0.0002)	L: p = 0.001; (M: W = 15, p = 0.0004)	na	na	L: p = 0.012; (M: W = 52.5, p = 0.0306)	na	L: p = 0.019; (M: W = 67.5, p = 0.0003)	L: p = 0.000; (M: W = 37.5, p = 0.0010)
Car1		-	L: p = 0.187; M: W = 47, p = 0.7201	L: p = 0.000; (M: W = 213, p = 0.0001)	L: p = 0.001; (M: W = 175, p = 0.0003)	L: p = 0.012; (M: W = 111, p = 0.0392)	L: p = 0.047; (M: W = 62.5, p = 0.0223)	L: p = 0.001; (M: W = 205, p = 0.3253)	L: p = 0.005; (M: W = 118, p = 0.1659)
Car2			-	L: p = 0.000; (M: W = 102, p = 0.0002)	L: p = 0.001; (M: W = 84, p = 0.0006)	L: p = 0.041; (M: W = 54.5, p = 0.0442)	L: p = 0.055; M: W = 30, p = 0.0294	L: p = 0.119; M: W = 118, p = 0.0656	L: p = 0.010; (M: W = 65, p = 0.0500)
Tar1				-	na	L: p = 0.007; (M: W = 59.5, p = 0.0214)	na	L: p = 0.012; (M: W = 76.5, p = 0.0001)	L: p = 0.000; (M: W = 42.5, p = 0.0005)
Tar2					-	L: p = 0.015; (M: W = 49, p = 0.0367)	na	L: p = 0.023; (M: W = 63, p = 0.0005)	L: p = 0.000; (M: W = 35, p = 0.0014)
Oma1						-	L: p = 0.151; M: W = 32.5, p = 0.2195	L: p = 0.362; M: W = 73, p = 0.0832	L: p = 0.496; M: W = 43.5, p = 0.2436
SAA1							-	L: p = 0.171; M: W = 22.5, p = 0.0264	L: p = 0.025; (M: W = 12.5, p = 0.0432)
SAA2								-	L: p = 0.169; M: W = 163, p = 0.376
SAA3									-

B2.1 Results from the statistical analyses (continued)

B2.1.1 Seedlings (continued)

	SAB1	SAB2	SAB3	Chu1	Chu2	Chu3	Chu4	Chu5	Chu6
Ba11	L: p = 0.001; (M: W = 120, p = 0.0993)	L: p = 0.000; (M: W = 104, p = 0.0270)	L: p = 0.001; (M: W = 112, p = 0.0520)	L: p = 0.000; (M: W = 0, p = 1.71e-06)	L: p = 0.000; (M: W = 0, p = 1.71e-06)	L: p = 0.004; (M: W = 16, p = 4.06e-05)	L: p = 0.000; (M: W = 8, p = 8.91e-06)	na	na
Bal2	L: p = 0.001; (M: W = 113, p = 0.1107)	L: p = 0.000; (M: W = 97.5, p = 0.0320)	L: p = 0.016; (M: W = 105, p = 0.0598)	L: p = 0.000; (M: W = 0, p = 3.05e-06)	L: p = 0.000; (M: W = 0, p = 3.05e-06)	L: p = 0.005; (M: W = 15, p = 6.42e-05)	L: p = 0.000; (M: W = 7.5, p = 1.50e-05)	na	na
Car1	L: p = 0.000; (M: W = 213, p = 0.0015)	L: p = 0.001; (M: W = 197, p = 0.0151)	L: p = 0.001; (M: W = 204, p = 0.0058)	L: p = 0.029; (M: W = 50, p = 0.1721)	L: p = 0.018; (M: W = 46, p = 0.1122)	L: p = 0.014; (M: W = 87.5, p = 0.4976)	L: p = 0.021; (M: W = 74, p = 0.9776)	L: p = 0.047; (M: W = 62.5, p = 0.0223)	L: p = 0.047; (M: W = 62.5, p = 0.0223)
Car2	L: p = 0.001; (M: W = 105, p = 0.0024)	L: p = 0.003; (M: W = 103, p = 0.0069)	L: p = 0.002; (M: W = 104, p = 0.0043)	L: p = 0.198; M: W = 26.5, p = 0.4318	L: p = 0.071; M: W = 14, p = 0.0445	L: p = 0.049; (M: W = 49, p = 0.1817)	L: p = 0.103; M: W = 47, p = 0.2582	L: p = 0.055; M: W = 30, p = 0.0294	L: p = 0.055; M: W = 30, p = 0.0294
Tar1	L: p = 0.000; (M: W = 128, p = 0.0892)	L: p = 0.000; (M: W = 111, p = 0.0227)	L: p = 0.001; (M: W = 119, p = 0.0453)	L: p = 0.000; (M: W = 0, p = 9.68e-07)	L: p = 0.000; (M: W = 0, p = 9.68e-07)	L: p = 0.003; (M: W = 17, p = 2.58e-05)	L: p = 0.000; (M: W = 8.5, p = 5.33e-06)	na	na
Tar2	L: p = 0.001; (M: W = 105, p = 0.1235)	L: p = 0.000; (M: W = 91, p = 0.0381)	L: p = 0.002; (M: W = 98, p = 0.0690)	L: p = 0.000; (M: W = 0, p = 5.46e-06)	L: p = 0.000; (M: W = 0, p = 5.46e-06)	L: p = 0.007; (M: W = 14, p = 0.0001)	L: p = 0.007; (M: W = 7, p = 2.53e-05)	na	na
Oma1	L: p = 0.043; (M: W = 103, p = 0.4216)	L: p = 0.743; M: W = 90.5, p = 1	L: p = 0.343; M: W = 97, p = 0.6812	L: p = 0.128; M: W = 3.5, p = 0.0004	L: p = 0.339; M: W = 0, p = 0.0001	L: p = 0.827; M: W = 24, p = 0.0402	L: p = 0.221; M: W = 17.5, p = 0.0118	L: p = 0.151; M: W = 32.5, p = 0.2195	L: p = 0.151; M: W = 32.5, p = 0.2195
SAA1	L: p = 0.051; M: W = 37.5, p = 0.3726	L: p = 0.032; (M: W = 32.5, p = 0.2152)	L: p = 0.056; M: W = 35, p = 0.2834	L: p = 0.024; M: W = 1.5, p = 0.0022)	L: p = 0.005; (M: W = 0, p = 0.0022)	L: p = 0.104; M: W = 5, p = 0.0109	L: p = 0.010; (M: W = 2.5, p = 0.0052)	na	na
SAA2	L: p = 0.025; W = 313, p = 0.0019)	L: p = 0.152; M: W = 278, p = 0.0433	L: p = 0.078; M: W = 292, p = 0.0143	L: p = 0.799; M: W = 17.5, p = 0.0001	L: p = 0.695; M: W = 10, p = 3.40e-05	L: p = 0.439; M: W = 115, p = 1	L: p = 0.872; M: W = 81.5, p = 0.1873	L: p = 0.171; M: W = 92.5, p = 0.0264	L: p = 0.171; M: W = 92.5, p = 0.0264
SAA3	L: p = 0.016; (M: W = 155, p = 0.0183)	L: p = 0.532; M: W = 134, p = 0.2213	L: p = 0.782; M: W = 144, p = 0.0802	L: p = 0.016; (M: W = 1.5, p = 0.0001)	L: p = 0.022; (M: W = 0, p = 7.52e-05)	L: p = 0.326; M: W = 45.5, p = 0.3353	L: p = 0.016; (M: W = 29, p = 0.0388)	L: p = 0.025; (M: W = 47.5, p = 0.0432)	L: p = 0.025; (M: W = 47.5, p = 0.0432)

B2.1 Results from the statistical analyses (continued)

B2.1.1 Seedlings (continued)

	SAB1	SAB2	SAB3	Chu1	Chu2	Chu3	Chu4	Chu5	Chu6
SAB1	-	L: p = 0.006; (M: W = 140, p = 0.3384)	L: p = 0.047; (M: W = 151, p = 0.6302)	L: p = 0.000; (M: W = 0, p = 3.08e-06)	L: p = 0.000; (M: W = 0, p = 3.08e-06)	L: p = 0.017; (M: W = 30, p = 0.0011)	L: p = 0.000; (M: W = 17, p = 0.0001)	L: p = 0.051; M: W = 52.5, p = 0.3726	L: p = 0.051; M: W = 52.5, p = 0.3726
SAB2	-		L: p = 0.338; (M: W = 173, p = 0.6619)	L: p = 0.015; (M: W = 3, p = 1.23e-05)	L: p = 0.061; M: W = 0, p = 6.06e-06)	L: p = 0.510; M: W = 49, p = 0.0340	L: p = 0.030; (M: W = 32.5, p = 0.0034)	L: p = 0.032; (M: W = 57.5, p = 0.2152)	L: p = 0.032; (M: W = 57.5, p = 0.2152)
SAB3			-	L: p = 0.003; (M: W = 2, p = 7.24e-06)	L: p = 0.007; (M: W = 0, p = 4.45e-06)	L: p = 0.196; M: W = 40, p = 0.0081	L: p = 0.004; (M: W = 25, p = 0.0007)	L: p = 0.056; M: W = 55, p = 0.2834	L: p = 0.056; M: W = 55, p = 0.2834
Chu1				-	L: p = 0.345; M: W = 20.5, p = 0.0272	L: p = 0.177; M: W = 95.5, p = 0.0006	L: p = 0.561; M: W = 89.5, p = 0.0031	L: p = 0.024; (M: W = 50, p = 0.0022)	L: p = 0.024; (M: W = 7.5, p = 0.0348)
Chu2					-	L: p = 0.478; M: W = 100, p = 0.0002	L: p = 0.661; M: W = 99.5, p = 0.0002	L: p = 0.005; (M: W = 50, p = 0.0022)	L: p = 0.005; (M: W = 50, p = 0.0022)
Chu3						-	L: p = 0.315; M: W = 34.5, p = 0.245	L: p = 0.104; M: W = 45, p = 0.0109	L: p = 0.104; M: W = 45, p = 0.0109
Chu4							-	L: p = 0.010; (M: W = 47.5, p = 0.0052)	L: p = 0.010; (M: W = 47.5, p = 0.0052)
Chu5								-	na

B2.1 Results from the statistical analyses (continued)

B2.1.2 Juveniles

	Bal2	Car1	Car2	Tar1	Tar2	Oma1	SAA1	SAA2	SAA3
Bal1	L: p = 0.655; M: W = 103, p = 0.4126	L: p = 0.004; (M: W = 40, p = 0.0008)	L: p = 0.000; (M: W = 2, p = 0.0001)	L: p = 0.022; (M: W = 49, p = 0.0010)	L: p = 0.000; (M: W = 64, p = 0.0253)	L: p = 0.711; M: W = 91, p = 0.4206	L: p = 0.079; M: W = 50, p = 0.2513	L: p = 0.018; (M: W = 48, p = 5.94e-05)	L: p = 0.006; (M: W = 58, p = 0.0474)
Bal2	-	L: p = 0.012; (M: W = 43, p = 0.0027)	L: p = 0.001; (M: W = 2, p = 0.0002)	L: p = 0.062; M: W = 56, p = 0.0051	L: p = 0.002; (M: W = 70, p = 0.0984)	L: p = 0.490; M: W = 95.5, p = 0.1587	L: p = 0.093; M: W = 52.5, p = 0.1144	L: p = 0.034; (M: W = 54, p = 0.0003)	L: p = 0.024; (M: W = 63, p = 0.1515)
Car1		-	L: p = 0.254; M: W = 17.5, p = 0.0140	L: p = 0.337; M: W = 150, p = 0.4098	L: p = 0.513; M: W = 118, p = 0.593	L: p = 0.012; (M: W = 127, p = 0.0024)	L: p = 0.020; (M: W = 68, p = 0.0074)	L: p = 0.706; M: W = 154, p = 0.5862	L: p = 0.489; M: W = 115, p = 0.2225
Car2			-	L: p = 0.040; (M: W = 106, p = 0.0032)	L: p = 0.029; (M: W = 84, p = 0.0089)	L: p = 0.001; (M: W = 69, p = 0.0005)	L: p = 0.005; (M: W = 35, p = 0.0042)	L: p = 0.579; M: W = 127, p = 0.0229	L: p = 0.063; M: W = 76.5, p = 0.0037
Tar1				-	L: p = 0.612; M: W = 119, p = 1	L: p = 0.040; (M: W = 144, p = 0.0020)	L: p = 0.031; (M: W = 78, p = 0.0052)	L: p = 0.260; M: W = 143, p = 0.1454	L: p = 0.797; M: W = 121, p = 0.4154
Tar2					-	L: p = 0.000; (M: W = 106, p = 0.0181)	L: p = 0.000; (M: W = 55, p = 0.0433)	L: p = 0.411; M: W = 130, p = 0.325	L: p = 0.828; M: W = 92, p = 0.6874
Oma1						-	L: p = 0.142; M: W = 27.5, p = 0.5716	L: p = 0.047; (M: W = 26, p = 0.0003)	L: p = 0.012; (M: W = 32, p = 0.0319)
SAA1							-	L: p = 0.079; M: W = 7.5, p = 0.0025	L: p = 0.009; (M: W = 13, p = 0.0445)
SAA2								-	L: p = 0.403; M: W = 189, p = 0.0750
SAA3									-

B2.1.2 Juveniles (continued)

	SAB1	SAB2	SAB3	Chu1	Chu2	Chu3	Chu4	Chu5	Chu6
Bal1	L: p = 0.000; (M: W = 89, p = 0.0329)	L: p = 0.000; (M: W = 57, p = 0.0013)	L: p = 0.000; (M: W = 50, p = 0.0006)	L: p = 0.000; (M: W = 0, p = 9.97e-06)	L: p = 0.001; (M: W = 0, p = 9.93e-06)	L: p = 0.037; (M: W = 2.5, p = 1.86e-05)	L: p = 0.001; (M: W = 0, p = 1e-05)	L: p = 0.019; (M: W = 15, p = 0.0176)	L: p = 0.011; (M: W = 2.5, p = 0.0007)
Bal2	L: p = 0.000; (M: W = 96, p = 0.1309)	L: p = 0.000; (M: W = 63, p = 0.0065)	L: p = 0.001; (M: W = 55, p = 0.0029)	L: p = 0.000; (M: W = 0, p = 2.12e-05)	L: p = 0.005; (M: W = 0, p = 2.11e-05)	L: p = 0.121; M: W = 4, p = 5.63e-05	L: p = 0.003; (M: W = 0, p = 2.13e-05)	L: p = 0.055; M: W = 16, p = 0.0466	L: p = 0.057; M: W = 3, p = 0.0017
Car1	L: p = 0.142; M: W = 158, p = 0.4058	L: p = 0.125; M: W = 121, p = 0.609	L: p = 0.405; M: W = 131, p = 0.8836	L: p = 0.045; (M: W = 9.5, p = 0.0003)	L: p = 0.836; M: W = 6.5, p = 0.0001	L: p = 0.228; M: W = 40.5, p = 0.0565	L: p = 0.688; M: W = 6.5, p = 0.0001	L: p = 0.829; M: W = 45, p = 0.5332	L: p = 0.576; M: W = 26.5, p = 0.3511
Car2	L: p = 0.976; M: W = 106, p = 0.0096	L: p = 0.963; M: W = 96, p = 0.0478	L: p = 0.579; M: W = 102, p = 0.0208	L: p = 0.500; M: W = 22.5, p = 0.2395	L: p = 0.342; M: W = 19.5, p = 0.1393	L: p = 0.024; (M: W = 53, p = 0.0951)	L: p = 0.429; M: W = 19, p = 0.1278	L: p = 0.294; M: W = 31.5, p = 0.0270	L: p = 0.138; M: W = 28, p = 0.1019
Tar1	L: p = 0.011; (M: W = 169, p = 0.6007)	L: p = 0.009; (M: W = 121, p = 0.2855)	L: p = 0.056; M: W = 126, p = 0.3671	L: p = 0.003; (M: W = 0, p = 9.12e-05)	L: p = 0.250; M: W = 30, p = 3.87e-05	L: p = 0.678; M: W = 30, p = 0.0058	L: p = 0.172; M: W = 3, p = 3.89e-05	L: p = 0.646; M: W = 43, p = 1	L: p = 0.911; M: W = 21, p = 0.0951
Tar2	L: p = 0.018; (M: W = 121, p = 0.8425)	L: p = 0.016; M: W = 87, p = 0.1293	L: p = 0.096; M: W = 99, p = 0.3053	L: p = 0.003; M: W = 1.5, p = 5.57e-05	L: p = 0.323; M: W = 0.5, p = 4.39e-05	L: p = 0.232; M: W = 23.5, p = 0.0062	L: p = 0.209; M: W = 0.5, p = 4.43e-05	L: p = 0.802; M: W = 34, p = 0.9619	L: p = 0.727; M: W = 19.5, p = 0.1536
Omal	L: p = 0.000; (M: W = 48, p = 0.0222)	L: p = 0.000; (M: W = 28, p = 0.0025)	L: p = 0.001; (M: W = 28, p = 0.0019)	L: p = 0.000; (M: W = 0, p = 8.51e-05)	L: p = 0.005; (M: W = 0, p = 8.45e-05)	L: p = 0.043; (M: W = 2, p = 0.0002)	L: p = 0.003; (M: W = 0, p = 8.57e-05)	L: p = 0.037; (M: W = 8.5, p = 0.0195)	L: p = 0.014; (M: W = 2, p = 0.0019)
SAA1	L: p = 0.001; (M: W = 20, p = 0.0434)	L: p = 0.002; (M: W = 13, p = 0.0127)	L: p = 0.004; (M: W = 10, p = 0.0079)	L: p = 0.002; (M: W = 0, p = 0.0022)	L: p = 0.009; (M: W = 0, p = 0.0021)	L: p = 0.015; (M: W = 0, p = 0.0021)	L: p = 0.007; (M: W = 0, p = 0.0022)	L: p = 0.042; (M: W = 2.5, p = 0.0248)	L: p = 0.003; (M: W = 0, p = 0.0073)
SAA2	L: p = 0.424; M: W = 254, p = 0.2153	L: p = 0.382; M: W = 194, p = 0.7308	L: p = 0.791; M: W = 214, p = 0.8736	L: p = 0.197; M: W = 16, p = 0.0002	L: p = 0.863; M: W = 16, p = 0.0001	L: p = 0.257; M: W = 65, p = 0.0508	L: p = 0.975; M: W = 16, p = 0.0001	L: p = 0.705; M: W = 73, p = 0.3631	L: p = 0.552; M: W = 45, p = 0.4672
SAA3	L: p = 0.029; (M: W = 101, p = 0.7565)	L: p = 0.026; (M: W = 73, p = 0.1359)	L: p = 0.117; M: W = 75.5, p = 0.1681	L: p = 0.009; (M: W = 4.5, p = 0.0003)	L: p = 0.353; M: W = 1.5, p = 0.0001	L: p = 0.478; M: W = 20, p = 0.0085	L: p = 0.252; M: W = 1.5, p = 0.0001	L: p = 0.764; M: W = 24.5, p = 0.5879	L: p = 0.929; M: W = 13, p = 0.0768

B2.1 Results from the statistical analyses (continued)

B2.1.2 Juveniles (continued)

	SAB1	SAB2	SAB3	Chu1	Chu2	Chu3	Chu4	Chu5	Chu6
SAB1	-	L: p = 0.917; M: W = 130, p = 0.2985	L: p = 0.482; M: W = 132, p = 0.3323	L: p = 0.356; M: W = 18, p = 0.0005	L: p = 0.254; M: W = 12, p = 0.0002	L: p = 0.010; (M: W = 52, p = 0.0644)	L: p = 0.344; M: W = 12, p = 0.0002	L: p = 0.219; M: W = 39.5, p = 0.6997	L: p = 0.106; M: W = 28.5, p = 0.2223
SAB2		-	L: p = 0.427; M: W = 172, p = 0.7613	L: p = 0.410; M: W = 24, p = 0.0016	L: p = 0.230; M: W = 17.5, p = 0.0005	L: p = 0.010; (M: W = 74, p = 0.4392)	L: p = 0.312; M: W = 17, p = 0.0005	L: p = 0.208; M: W = 52.5, p = 0.5974	L: p = 0.102; M: W = 41.5, p = 0.8217
SAB3			-	L: p = 0.141; M: W = 17.5, p = 0.0005	L: p = 0.581; M: W = 11.5, p = 0.0002	L: p = 0.043; (M: W = 60, p = 0.1473)	L: p = 0.728; M: W = 11, p = 0.0002	L: p = 0.425; M: W = 52.5, p = 0.5976	L: p = 0.238; M: W = 36, p = 0.5228
Chu1				-	L: p = 0.088; M: W = 50, p = 1	L: p = 0.004; (M: W = 89, p = 0.0031)	L: p = 0.120; M: W = 48, p = 0.909	L: p = 0.114; M: W = 46.5, p = 0.0097	L: p = 0.052; M: W = 45.5, p = 0.0134
Chu2					-	L: p = 0.142; M: W = 96, p = 0.0005	L: p = 0.849; M: W = 48, p = 0.9084	L: p = 0.705; M: W = 48.5, p = 0.0046	L: p = 0.434; M: W = 48.5, p = 0.0046
Chu3						-	L: p = 0.096; M: W = 4, p = 0.0005	L: p = 0.429; M: W = 38.5, p = 0.1062	L: p = 0.588; M: W = 30.5, p = 0.5317
Chu4							-	L: p = 0.599; M: W = 48.5, p = 0.00467	L: p = 0.344; M: W = 48.5, p = 0.0047
Chu5								-	L: p = 0.756; M: W = 7, p = 0.2812

B2.1 Results from the statistical analyses (continued)

B2.1.3 Adults

	Bal2	Car1	Car2	Tar1	Tar2	Oma1	SAA1	SAA2	SAA3
Bal1	L: p = 0.420; M: W = 76, p = 0.0818	L: p = 0.001; (M: W = 8.5, p = 1.07e-05)	L: p = 0.001; (M: W = 0, p = 0.0002)	L: p = 0.827; M: W = 99, p = 0.1837	L: p = 0.236; M: W = 47, p = 0.0069	L: p = 0.012; (M: W = 137, p = 0.0020)	L: p = 0.046; (M: W = 70, p = 0.0123)	L: p = 0.157; M: W = 162, p = 0.5201	L: p = 0.099; M: W = 56, p = 0.0635
Bal2	-	L: p = 0.001; (M: W = 18.5, p = 0.0001)	L: p = 0.002; (M: W = 0, p = 0.0002)	L: p = 0.252; M: W = 142, p = 0.5941	L: p = 0.648; M: W = 75.5, p = 0.2033	L: p = 0.001; (M: W = 146, p = 7.48e-05)	L: p = 0.008; (M: W = 74, p = 0.0015)	L: p = 0.342; M: W = 213, p = 0.2245	L: p = 0.261; M: W = 75.5, p = 0.4922
Car1		-	L: p = 0.888; M: W = 12, p = 0.0047	L: p = 0.000; (M: W = 245, p = 9.55e-06)	L: p = 0.003; (M: W = 180, p = 0.0011)	L: p = 0.001; (M: W = 150, p = 2.61e-05)	L: p = 0.014; (M: W = 75, p = 0.0012)	L: p = 0.003; (M: W = 317, p = 1.51e-05)	L: p = 0.015; (M: W = 153, p = 0.0024)
Car2			-	L: p = 0.000; (M: W = 119, p = 0.0002)	L: p = 0.004; (M: W = 98, p = 0.0003)	L: p = 0.002; (M: W = 70, p = 0.0004)	L: p = 0.024; (M: W = 35, p = 0.0049)	L: p = 0.008; (M: W = 160, p = 8.80e-05)	L: p = 0.020; (M: W = 84, p = 0.0004)
Tar1				-	L: p = 0.123; M: W = 71.5, p = 0.0608	L: p = 0.004; (M: W = 160, p = 0.0001)	L: p = 0.020; (M: W = 81, p = 0.0026)	L: p = 0.110; M: W = 223, p = 0.4533	L: p = 0.053; M: W = 70.5, p = 0.1676
Tar2					-	L: p = 0.001; (M: W = 138, p = 5.55e-05)	L: p = 0.008; (M: W = 69.5, p = 0.0015)	L: p = 0.508; M: W = 232, p = 0.0257	L: p = 0.450; M: W = 88, p = 0.8567
Oma1						-	L: p = 0.479; M: W = 25.5, p = 1	L: p = 0.022; (M: W = 17.5, p = 9.96e-05)	L: p = 0.003; (M: W = 13, p = 0.0013)
SAA1							-	L: p = 0.086; M: W = 5.5, p = 0.0016	L: p = 0.027; (M: W = 6, p = 0.0112)
SAA2								-	L: p = 0.967; M: W = 97.5, p = 0.1585
SAA3									-

211

B2.1 Results from the statistical analyses (continued)

B2.1.3 Adults (continued)

	SAB1	SAB2	SAB3	Chu1	Chu2	Chu3	Chu4	Chu5	Chu6
Bal1	L: p=0.730; M: W=110, p=0.243	L: p=0.067; M: W=95, p=0.0917	L: p=0.404; M: W=123, p=0.4617	L: p=0.894; M: W=25.5, p=0.0041	L: p=0.186; M: W=65.5, p=0.4493	L: p=0.065; M: W=0, p=2.52e-05	L: p=0.548; M: W=3, p=4.97e-05	L: p=0.053; M: W=4.5, p=0.0035	L: p=0.380; M: W=3, p=0.0023
Bal2	L: p=0.664; M: W=149, p=0.6363	L: p=0.194; M: W=133, p=0.942	L: p=0.796; M: W=163, p=0.3159	L: p=0.399; M: W=46.5, p=0.1172	L: p=0.028; (M: W=95, p=0.2722)	L: p=0.239; M: W=0, p=3.43e-05	L: p=0.148; M: W=8, p=0.0002	L: p=0.126; M: W=9, p=0.0141	L: p=0.122; M: W=7, p=0.0083
Car1	L: p=0.000; (M: W=255, p=1.61e-05)	L: p=0.005; (M: W=239, p=0.0002)	L: p=0.001; (M: W=255, p=1.46e-05)	L: p=0.005; (M: W=133, p=0.0014)	L: p=0.002; (M: W=148.5, p=4.85e-05)	L: p=0.015; (M: W=61.5, p=0.46686)	L: p=0.003; (M: W=91.5, p=0.373)	L: p=0.116; M: W=51, p=0.2551	L: p=0.026; (M: W=49, p=0.3329)
Car2	L: p=0.001; (M: W=126, p=0.0002)	L: p=0.009; (M: W=126, p=0.0002)	L: p=0.003; (M: W=126, p=0.0001)	L: p=0.007; (M: W=70, p=0.0007)	L: p=0.003; (M: W=70, p=0.0007)	L: p=0.017; (M: W=68.5, p=0.0012)	L: p=0.005; (M: W=70, p=0.0007)	L: p=0.129; M: W=35, p=0.0056	L: p=0.038; M: W=35, p=0.0055
Tar1	L: p=0.553; M: W=154, p=1	L: p=0.037; (M: W=134, p=0.5282)	L: p=0.296; M: W=172, p=0.5489	L: p=0.969; M: W=41.5, p=0.0291	L: p=0.166; M: W=101, p=0.4452	L: p=0.016; (M: W=0, p=2.10e-05)	L: p=0.619; M: W=2.5, p=3.64e-05	L: p=0.015; (M: W=6.5, p=0.0052)	L: p=0.357; M: W=2, p=0.0017
Tar2	L: p=0.403; M: W=173, p=0.0790	L: p=0.346; M: W=152, p=0.3403	L: p=0.940; M: W=182, p=0.0339	L: p=0.249; M: W=67, p=0.8828	L: p=0.017; (M: W=110, p=0.0213)	L: p=0.530; (M: W=0, p=4.57e-05)	L: p=0.084; M: W=17.5, p=0.0022	L: p=0.270; M: W=13.5, p=0.05013	L: p=0.090; M: W=13.5, p=0.05013
Oma1	L: p=0.009; (M: W=20.5, p=0.0006)	L: p=0.004; (M: W=11.5, p=0.0001)	L: p=0.025; (M: W=26, p=0.0015)	L: p=0.024; (M: W=0, p=0.0001)	L: p=0.039; (M: W=3.5, p=0.0003)	L: p=0.000; (M: W=0, p=0.0001)	L: p=0.006; (M: W=0, p=0.0001)	L: p=0.000; (M: W=0, p=0.0011)	L: p=0.032; (M: W=0, p=0.0011)
SAA1	L: p=0.039; (M: W=9.5, p=0.0080)	L: p=0.028; (M: W=4.5, p=0.0026)	L: p=0.087; M: W=12, p=0.0130	L: p=0.072; M: W=0, p=0.0023	L: p=0.060; M: W=0.5, p=0.0027	L: p=0.000; (M: W=0, p=0.0024)	L: p=0.021; (M: W=0, p=0.0023)	L: p=0.002; M: W=0, p=0.0095	L: p=0.046; (M: W=0, p=0.0092)
SAA2	L: p=0.205; M: W=186, p=0.586	L: p=0.901; M: W=165, p=0.2642	L: p=0.467; M: W=201, p=0.8728	L: p=0.227; M: W=53.5, p=0.0155	L: p=0.077; M: W=107, p=0.7623	L: p=0.786; M: W=12.5, p=5.48e-05	L: p=0.149; M: W=23, p=0.0003	L: p=0.878; M: W=14.5, p=0.0096	L: p=0.222; M: W=13, p=0.0073
SAA3	L: p=0.155; M: W=140, p=0.1799	L: p=0.862; M: W=121, p=0.6102	L: p=0.482; M: W=143, p=0.1463	L: p=0.144; M: W=61, p=0.9735	L: p=0.024; (M: W=84, p=0.117)	L: p=0.756; M: W=4.5, p=0.0003	L: p=0.066; M: W=17.5, p=0.0054	L: p=0.808; M: W=15, p=0.1243	L: p=0.113; M: W=11, p=0.0496

B2.1 Results from the statistical analyses (continued)

B2.1.3 Adults (continued)

	SAB1	SAB2	SAB3	Chu1	Chu2	Chu3	Chu4	Chu5	Chu6
SAB1	-	L: p = 0.101; M: W = 141, p = 0.5145	L: p = 0.554; M: W = 181, p = 0.5654	L: p = 0.673; M: W = 46.5, p = 0.0371	L: p = 0.121; M: W = 106, p = 0.4675	L: p = 0.153; M: W = 0, p = 1.69e-05	L: p = 0.375; M: W = 6.5, p = 6.56e-05	L: p = 0.103; M: W = 7.5, p = 0.0056	L: p = 0.295; M: W = 6, p = 0.0040
SAB2		-	L: p = 0.346; M: W = 198, p = 0.2652	L: p = 0.114; M: W = 65.5, p = 0.2472	L: p = 0.022; (M: W = 113, p = 0.2874)	L: p = 0.634; M: W = 7.5, p = 8.18e-05	L: p = 0.057; M: W = 25.5, p = 0.0021	L: p = 0.924; M: W = 18, p = 0.0476	L: p = 0.109; M: W = 14, p = 0.0224
SAB3			-	L: p = 0.438; M: W = 39.5, p = 0.0157	L: p = 0.124; M: W = 93, p = 0.903	L: p = 0.631; M: W = 5, p = 4.74e-05	L: p = 0.273; M: W = 12, p = 0.0002	L: p = 0.432; M: W = 10, p = 0.0096	L: p = 0.296; M: W = 6, p = 0.0038
Chu1				-	L: p = 0.272; M: W = 88, p = 0.0041	L: p = 0.067; M: W = 0, p = 0.0002	L: p = 0.677; M: W = 9.5, p = 0.0023	L: p = 0.066; M: W = 8, p = 0.0409	L: p = 0.468; M: W = 7.5, p = 0.0348
Chu2					-	L: p = 0.000; (M: W = 0, p = 0.0002)	L: p = 0.342; M: W = 0, p = 0.0002	L: p = 0.001; (M: W = 1.5, p = 0.0044)	L: p = 0.874; M: W = 0, p = 0.0024
Chu3						-	L: p = 0.004; (M: W = 98, p = 0.0003)	L: p = 0.375; M: W = 43.5, p = 0.0254	L: p = 0.004; (M: W = 50, p = 0.0025)
Chu4							-	L: p = 0.008; (M: W = 21.5, p = 0.7093)	L: p = 0.533; M: W = 34.5, p = 0.2594
Chu5								-	L: p = 0.016; (M: W = 15, p = 0.6733)

B2.1 Results from the statistical analyses (continued)

B2.1.4 Abundance

	Bal2	Car1	Car2	Tar1	Tar2	Oma1	SAA1	SAA2	SAA3
Bal1	L: p = 0.388; M: W = 81, p = 0.1253	L: p = 0.000; (M: W = 5.5, p = 6.32e-06)	L: p = 0.001; (M: W = 0, p = 0.0002)	L: p = 0.462; M: W = 73.5, p = 0.0248	L: p = 0.047; (M: W = 39.5, p = 0.0027)	L: p = 0.974; M: W = 123, p = 0.0215	L: p = 0.024; (M: W = 70, p = 0.0127)	L: p = 0.041; (M: W = 67, p = 0.0009)	L: p = 0.37; M: W = 45.5, p = 0.0197
Bal2	-	L: p = 0.001; (M: W = 17, p = 7.83e-05)	L: p = 0.002; (M: W = 0, p = 0.0002)	L: p = 0.976; M: W = 108, p = 0.4701	L: p = 0.247; M: W = 63, p = 0.0687	L: p = 0.550; M: W = 133, p = 0.0014	L: p = 0.016; (M: W = 74, p = 0.0015)	L: p = 0.093; M: W = 99.5, p = 0.0299	L: p = 0.116; M: W = 63.5, p = 0.2016
Car1		-	L: p = 0.006; (M: W = 70.5, p = 0.8241)	L: p = 0.000; (M: W = 240, p = 2.45e-05)	L: p = 0.002; (M: W = 176, p = 0.0022)	L: p = 0.005; (M: W = 144, p = 0.0001)	L: p = 0.027; (M: W = 75, p = 0.0012)	L: p = 0.001; (M: W = 265, p = 0.0062)	L: p = 0.008; (M: W = 144, p = 0.0095)
Car2			-	L: p = 0.001; (M: W = 119, p = 0.0002)	L: p = 0.006; (M: W = 98, p = 0.0003)	L: p = 0.010; (M: W = 70, p = 0.0006)	L: p = 0.040; (M: W = 35, p = 0.0049)	L: p = 0.017; (M: W = 158, p = 0.0002)	L: p = 0.025; (M: W = 83, p = 0.0006)
Tar1				-	L: p = 0.329; M: W = 84, p = 0.169	L: p = 0.586; M: W = 147, p = 0.0018	L: p = 0.070; M: W = 81.5, p = 0.0023	L: p = 0.085; M: W = 123, p = 0.0467	L: p = 0.135; M: W = 76, p = 0.2572
Tar2					-	L: p = 0.167; M: W = 128, p = 0.0007	L: p = 0.007; (M: W = 69.5, p = 0.0015)	L: p = 0.261; M: W = 144, p = 0.5934	L: p = 0.412; M: W = 83, p = 0.9794
Oma1						-	L: p = 0.208; M: W = 30.5, p = 0.4602	L: p = 0.109; M: W = 20.5, p = 0.0002	L: p = 0.121; M: W = 17.5, p = 0.0047
SAA1							-	L: p = 0.085; M: W = 1, p = 0.0007	L: p = 0.046; (M: W = 3.5, p = 0.0053)
SAA2								-	L: p = 0.653; M: W = 153, p = 0.6254
SAA3									-

B2.1 Results from the statistical analyses (continued)

B2.1.4 Abundance (continued)

	SAB1	SAB2	SAB3	Chu1	Chu2	Chu3	Chu4	Chu5	Chu6
Bal1	L: p = 0.005; (M: W = 96, p = 0.0993)	L: p = 0.001; (M: W = 69, p = 0.0098)	L: p = 0.003; (M: W = 89.5, p = 0.0609)	L: p = 0.005; (M: W = 0, p = 2.67e-05)	L: p = 0.062; M: W = 0, p = 2.63e-05	L: p = 0.448; M: W = 0, p = 2.65e-05	L: p = 0.090; M: W = 0, p = 2.65e-05	L: p = 0.109; M: W = 2, p = 0.0019	L: p = 0.325; M: W = 0, p = 0.0010
Bal2	L: p = 0.031; (M: W = 123, p = 0.6763)	L: p = 0.005; (M: W = 88, p = 0.0915)	L: p = 0.016; (M: W = 113, p = 0.4241)	L: p = 0.049; (M: W = 0, p = 3.45e-05)	L: p = 0.278; M: W = 0, p = 3.42e-05	L: p = 0.970; M: W = 0, p = 3.43e-05	L: p = 0.300; M: W = 0, p = 3.43e-05	L: p = 0.294; M: W = 7.5, p = 0.0096	L: p = 0.165; M: W = 2, p = 0.0021
Car1	L: p = 0.001; (M: W = 235, p = 0.0003)	L: p = 0.003; (M: W = 204, p = 0.0132)	L: p = 0.002; (M: W = 225, p = 0.0012)	L: p = 0.012; (M: W = 61, p = 0.4535)	L: p = 0.009; (M: W = 60, p = 0.4198)	L: p = 0.006; (M: W = 70.5, p = 0.8241)	L: p = 0.009; (M: W = 66, p = 0.6367)	L: p = 0.067; M: W = 52, p = 0.2196	L: p = 0.034; (M: W = 45, p = 0.5401)
Car2	L: p = 0.007; (M: W = 126, p = 0.0002)	L: p = 0.017; (M: W = 124, p = 0.0002)	L: p = 0.009; (M: W = 125, p = 0.0002)	L: p = 0.027; (M: W = 62, p = 0.0096)	L: p = 0.019; (M: W = 61.5, p = 0.0109)	L: p = 0.012; (M: W = 64, p = 0.0053)	L: p = 0.022; (M: W = 63, p = 0.0072)	L: p = 0.114; M: W = 34.5, p = 0.0073)	L: p = 0.053; M: W = 35, p = 0.0056
Tar1	L: p = 0.040; (M: W = 158, p = 0.8945)	L: p = 0.006; (M: W = 108, p = 0.1364)	L: p = 0.021; (M: W = 143, p = 0.74)	L: p = 0.102; M: W = 0, p = 2.13e-05	L: p = 0.376; M: W = 0, p = 2.10e-05	L: p = 0.955; M: W = 0.5, p = 2.36e-05	L: p = 0.370; M: W = 1, p = 2.64e-05	L: p = 0.378; M: W = 10, p = 0.0117	L: p = 0.292; M: W = 7, p = 0.0059
Tar2	L: p = 0.217; (M: W = 155, p = 0.2865)	L: p = 0.048; (M: W = 114, p = 0.6477)	L: p = 0.129; (M: W = 142, p = 0.5551)	L: p = 0.397; M: W = 0, p = 4.66e-05	L: p = 0.985; M: W = 0, p = 4.59e-05	L: p = 0.288; M: W = 0, p = 4.62e-05	L: p = 0.910; M: W = 0.5, p = 5.22e-05	L: p = 0.758; M: W = 18.5, p = 0.1368	L: p = 0.054; M: W = 12.5, p = 0.0397
Oma1	L: p = 0.036; (M: W = 29, p = 0.0032)	L: p = 0.011; (M: W = 22, p = 0.0010)	L: p = 0.023; (M: W = 32, p = 0.0050)	L: p = 0.060; M: W = 0, p = 0.0001	L: p = 0.208; M: W = 0, p = 0.0001	L: p = 0.616; M: W = 0, p = 0.0001	L: p = 0.235; M: W = 0, p = 0.0001	L: p = 0.283; M: W = 2.5, p = 0.0052	L: p = 0.591; M: W = 2, p = 0.0042
SAA1	L: p = 0.009; (M: W = 7, p = 0.0047)	L: p = 0.005; (M: W = 4.5, p = 0.0027)	L: p = 0.007; (M: W = 9, p = 0.0072)	L: p = 0.002; (M: W = 0, p = 0.0025)	L: p = 0.009; (M: W = 0, p = 0.0024)	L: p = 0.017; (M: W = 0, p = 0.0024)	L: p = 0.031; (M: W = 0, p = 0.0024)	L: p = 0.061; M: W = 0, p = 0.0097	L: p = 0.089; M: W = 0, p = 0.0092
SAA2	L: p = 0.677; M: W = 261, p = 0.1592	L: p = 0.808; M: W = 204, p = 0.937	L: p = 0.847; M: W = 244, p = 0.3432	L: p = 0.558; M: W = 16, p = 0.0001	L: p = 0.343; M: W = 16, p = 0.0001	L: p = 0.163; M: W = 20, p = 0.0002	L: p = 0.377; M: W = 18, p = 0.0002	L: p = 0.597; M: W = 40, p = 0.3067	L: p = 0.158; M: W = 32.5, p = 0.0516
SAA3	L: p = 0.871; M: W = 127, p = 0.445	L: p = 0.384; M: W = 95, p = 0.5959	L: p = 0.678; M: W = 117, p = 0.734	L: p = 0.843; M: W = 5, p = 0.0003	L: p = 0.483; M: W = 3, p = 0.0002	L: p = 0.179; M: W = 6.5, p = 0.0005	L: p = 0.551; M: W = 6, p = 0.0004	L: p = 0.767; M: W = 19, p = 0.2663	L: p = 0.117; M: W = 11, p = 0.0498

B2.1 Results from the statistical analyses (continued)

B2.1.4 Abundance (continued)

	SAB1	SAB2	SAB3	Chu1	Chu2	Chu3	Chu4	Chu5	Chu6
SAB1	-	L: p = 0.370; M: W = 123, p = 0.2219	L: p = 0.750; M: W = 154, p = 0.7994	L: p = 0.660; M: W = 3, p = 3.31e-05	L: p = 0.288; M: W = 2, p = 2.65e-05	L: p = 0.063; M: W = 4.5, p = 4.51e-05	L: p = 0.357; M: W = 4, p = 4.04e-05	L: p = 0.614; M: W = 19, p = 0.0569	L: p = 0.036; (M: W = 17, p = 0.0399)
SAB2		-	L: p = 0.556; M: W = 193, p = 0.3414	L: p = 0.236; M: W = 11.5, p = 0.0002	L: p = 0.088; M: W = 7.5, p = 8.23e-05	L: p = 0.017; (M: W = 13.5, p = 0.0003)	L: p = 0.118; M: W = 12, p = 0.0002	L: p = 0.320; M: W = 35, p = 0.478	L: p = 0.019; (M: W = 30, p = 0.2781)
SAB3			-	L: p = 0.473; M: W = 5, p = 4.97e-05	L: p = 0.190; M: W = 4.5, p = 4.38e-05	L: p = 0.039; (M: W = 8, p = 9.04e-05)	L: p = 0.245; M: W = 7.5, p = 8.20e-05	L: p = 0.491; M: W = 24.5, p = 0.135	L: p = 0.027; (M: W = 19, p = 0.0564)
Chu1				-	L: p = 0.447; M: W = 47, p = 0.8495	L: p = 0.070; M: W = 64, p = 0.3062	L: p = 0.572; M: W = 55.5, p = 0.7046	L: p = 0.813; M: W = 46.5, p = 0.0101	L: p = 0.014; (M: W = 50, p = 0.0026)
Chu2					-	L: p = 0.311; M: W = 71, p = 0.119	L: p = 0.931; M: W = 62.5, p = 0.3613	L: p = 0.790; M: W = 47.5, p = 0.0068	L: p = 0.063; M: W = 50, p = 0.0026
Chu3						-	L: p = 0.350; M: W = 43, p = 0.6213	L: p = 0.342; M: W = 46, p = 0.0117	L: p = 0.175; M: W = 50, p = 0.0026
Chu4							-	L: p = 0.863; M: W = 46, p = 0.0119	L: p = 0.122; M: W = 49.5, p = 0.0032
Chu5								-	L: p = 0.165; M: W = 10, p = 0.6714

B2.2 Results from the statistical analysis of the pairwise tested (t-tests) experimental set up data (counted number of successfully germinated seeds) from the experiment on germination success (U – untrested, P – pre-treated as described in the Materials & Methods-section; 0, 5, and 10 cm correspond to burial depth)

	U - 5 cm	U - 10 cm	P - 0 cm	P - 5 cm	P - 10 cm
U - 0 cm	$p < 0.001$, $t = -9.823$, $df = 15.067$	$p < 0.001$, $t = -11.339$, $df = 13.755$	$p = 0.067$, $t = 1.952$, $df = 17.589$	$p < 0.001$, $t = -11.015$, $df = 12.926$	$p < 0.001$, $t = -10.791$, $df = 14.311$
U - 5 cm	-	$p = 0.135$, $t = -1.567$, $df = 17.596$	$p < 0.001$, $t = 13.349$, $df = 16.430$	$p = 0.425$, $t = -0.818$, $df = 16.891$	$p = 0.324$, $t = -1.014$, $df = 17.869$
U - 10 cm		-	$p < 0.001$, $t = 15.274$, $df = 15.080$	$p = 0.382$, $t = 0.896$, $df = 17.789$	$p = 0.591$, $t = 0.548$, $df = 17.920$
P - 0 cm			-	$p < 0.001$, $t = -15.057$, $df = 14.125$	$p < 0.001$, $t = -14.561$, $df = 15.680$
P - 5 cm				-	$p = 0.777$, $t = -0.287$, $df = 17.469$

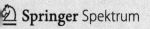